C0-CFB-860

The Planned Economies of Eastern Europe

ALAN H. SMITH

HOLMES & MEIER PUBLISHERS, INC.
New York

First Published in the United States of America 1983 by
Holmes & Meier Publishers, Inc.,
30 Irving Place,
New York, N.Y. 10003

Library of Congress Cataloging in Publication Data

Smith, Alan H.
The planned economies of Eastern Europe.

Includes bibliographical references and index.
1. Europe, Eastern--Economic policy. 2. Monetary policy--Europe, Eastern. 3. Europe, Eastern--Foreign economic relations. I. Title.
HC244.S553 1983 338.947 83-10748
ISBN 0-8419-0891-5

Printed and bound in Great Britain

CONTENTS

TABLES

PREFACE

Following a period of impressive industrial growth, combined with low rates of open inflation and unemployment, the performance of the East European economies in the 1980s is a cause of considerable concern. Throughout Eastern Europe industrial growth has declined to the point where the prospect of zero (or even negative) growth has arisen. In all countries substantial increases in retail prices have been announced while the volume of household money savings has approached (or reached) the equivalent of a year's turnover in state retail stores; the level of indebtedness to the West exceeds safe limits for many countries; investment in CMEA (Council for Mutual Economic Assistance) projects to develop energy and raw material sources situated in the USSR is putting immediate strains on current consumption levels; reports of labour unrest, consumer dissatisfaction, black market activity, petty and large-scale corruption frequently reach the West; and the need to allocate labour more efficiently may give rise to frictional unemployment. Although the underlying problems are common to each of the East European countries their scale varies considerably.

The purpose of this book is to examine the adaptation of the Soviet planning system to East European circumstances, paying particular attention to the problems of economic relations with the USSR and the problems of foreign trade, the constraints to domestic policies established by relations with the USSR and the problems of domestic consumer equilibrium.

It is hoped that the book will be of some value to readers with a general interest in Eastern Europe, readers with an interest in economic systems and readers with a knowledge of the Soviet system wishing to extend that knowledge to Eastern Europe. I have, however, assumed no prior knowledge of the Soviet economic system and each section contains a brief outline of the original conceptions of Soviet economists in the 1920s, the development of the system in the USSR and its subsequent application in Eastern Europe. I have tried to simplify the economic analysis for the benefit of non-economists (although some of the theoretical sections are somewhat technical) and readers wishing greater technical detail may pursue the references in the text and the bibliography.

Preface

A reference list of official Soviet and East European statistical publications from which quantitative statements have been derived is provided after the bibliography. The use of western data sources is indicated in the text. The term billion always refers to an American billion.

A survey of this nature clearly depends very heavily on other people's work and ideas. The reference list indicates many, but not all of my debts. I am also very grateful to both speakers and participants in seminars throughout the United Kingdom (but particularly at the London School of Economics) for ideas that it is impossible to attribute accurately. In particular, however, I owe a considerable debt to Professor R. W. Davies for first stimulating my interest in this area and subsequently to the ideas of Michael Ellman, Staszek Gomulka, Philip Hanson, Michael Kaser, Colin Lawson, Alec Nove, Mario Nuti, Richard Portes and Peter Wiles, amongst many others. I have also benefited from the opportunity to present papers, on which this book has been based, at the LSE, Chatham House, St Antony's College, Oxford, the Universities of Wales, Reading and Bath, and Dartmouth College, New Hampshire and I thank participants for their comments. I remain responsible for any errors and misinterpretations.

Finally, I would like to thank my wife, Ruth, who typed the whole manuscript and without whose help the book would never have been completed.

PART ONE

THE ORIGINS AND OPERATION OF CENTRAL PLANNING

1 THE DEVELOPMENT OF THE STALINIST ECONOMIC SYSTEM IN THE USSR

Introduction

It is tempting for the economist to depict the Soviet model of a centrally-planned economy as a pure economic system that operates with a number of imperfections that arise from the nature of central planning itself.

In fact the Soviet system of planning emerged as a compromise between two vastly different conceptions of the nature of socialist economic organisation. Many theorists saw the launching of the first five-year plan and the collectivisation of agriculture in 1929 as an initial irreversible move towards the installation of a full communist, centrally-directed, moneyless economy implying the defeat of those schools who wished to guide economic activity through controls over the money and banking system, while retaining considerable elements of the market and small-scale private industry and agriculture (Lewin, 1975, p. 97).

The legal recognition of the need to maintain private production and trade in grain in 1932 reversed the move to a moneyless economy and subsequently a compromise role was accorded to the use of monetary relations, not just between private production and trade, but between state economic organisations themselves. A number of practical refinements have been made, but the system remains effectively unaltered fifty years later. Money remains subservient to the planned control of the economy, enterprises respond to central instructions not monetary criteria, but their progress is checked by monetary accounts. Workers are paid in money and spend this on what is made available at centrally-determined prices in state stores. Beneath this a whole framework of legal and (normally tolerated, occasionally suppressed) illegal market activities take place.

Despite the fact that the system emerged as a compromise to prevailing conditions in the USSR in the 1930s, it was imitated from 1948 onwards in a virtually unchanged form in the East European countries that came under Soviet hegemony.

Pre-Stalinist Economic Systems in the Soviet Union

The Bolshevik leadership was faced with a critical economic dilemma immediately on taking power. Capitalism had been viewed as an inevitable stage of development in which the foundations of an industrial society would be established. The Revolution however had taken place in what was predominantly an agricultural society. Furthermore, although Marx left no detailed guidelines for the operation of a socialist economy, there were strong centralist tendencies in both Marxian and Bolshevik economic analysis, while the events of 1917 were inherently centrifugal not centripetal. Disagreement therefore arose on such issues as whether the bourgeois stage of development could be omitted, on the nature of economic organisation and on the relationship between the industrial proletariat and the peasantry.

Lenin's initial conception of industrial organisation was determined by the desire to establish large-scale industry subject to central direction and was influenced by the experience of German war-time planning and by contemporary analysis of the role of financial capital in the process of economic development. He envisaged a system that could be described as State Capitalism, involving the emergence of a monopolistic industry, possibly remaining in private hands but subject to central financial control over investment and subject to operational checks by the emerging workers' committees.

The operation of this concept of planning, which involved the nationalisation of banks but retained private industry, was prevented by the lack of co-operation by private owners, while fears of the wholesale transfer of industrial capital into German hands strengthened the position of those who favoured a policy of the planned nationalisation of industry. Finally, the Civil War (1918–21) and the consequent resort to the printing press as a source of finance brought about the destruction of the monetary system and led to the emergence of an economic mechanism that was subsequently described as 'War Communism'. The major features of industrial organisation under War Communism were the nationalisation of all large-scale industry, the establishment of a moneyless economy involving in the first instance the moneyless transfers of inputs and outputs between enterprises and the establishment of a planning bureaucracy in the shape of the Supreme Economic Council (VSNKH), which functioned not as initially conceived to oversee the financial resources of the government, but to effect transfers of inputs and outputs between enterprises on the basis of central instructions. The progressive collapse of the monetary system led to the increased role of non-monetary relations between enterprises and workers, the establishment of payment either in kind and/or in basic commodities, the development of a barter economy and the free provision of infrastructural goods such as transportation and communications. In the absence of monetary incentives, labour was partially mobilised by moral appeals and partially by a system of coercion which ultimately culminated in the militarisation of labour. In addition, physical wealth was frequently confiscated and redistributed.

A critical feature of War Communism was the process of exchange between agriculture and industry. As a result of the collapse of the monetary system the peasant had little or no incentive to produce for the market. Although attempts were made to send industrial commodities to the countryside and exchange them for agricultural products the system was too unreliable and inefficient to generate an agricultural surplus that would be sufficient to meet the demands of the town. The Bolshevik response was to requisition all foodstuffs above that required by the peasant for subsistence and seed, a measure enforced locally by Committees of the Village Poor. The inevitable consequence was that the peasant reduced the sown area to the level sufficient to meet his own needs, resulting in further reductions in output, leading to further measures of coercion, which threatened to create a permanent break between the Bolshevik regime and the rural population.

There was some disagreement amongst the Bolshevik leadership during this period between those who saw the system of War Communism as an initial, if somewhat premature, development of a communist economy and those who regarded the system purely as an undesirable consequence of the inflationary pressures and chaos engendered by the Civil War.

Those holding the second view proposed to re-establish cooperative relations between the town and country by replacing compulsory requisitioning with a tax in kind and permitting the sale of any produce above this amount on the market. This in turn required the re-establishment of a stable currency and a degree of market orientation on behalf of state producers.

This policy was implemented by a series of measures enacted between 1921 and 1923, known as the New Economic Policy

(NEP). The major significance of NEP for current debates on the nature of the economic system under socialism is that its supporters saw it as a viable socialist alternative to the economic system proposed by Stalin, whilst its opponents argued that it was not socialist in content, that it depended too heavily on the support of capitalist traders, and that it would not have stimulated a sufficient rate of industrial growth to defend the country from outside attack. NEP was, in essence, a system of market socialism in which nationalised large-scale industry coexisted with small-scale private industry. The state sector of industry was largely composed of commercially autonomous trusts that determined their own levels of output and prices on the basis of market criteria, rather than operating according to physical instructions from the centre concerning the supply of inputs and the destination of outputs as subsequently prevailed in the Stalinist centrally-planned economy.

One feature that emerged in NEP that was subsequently retained in the Stalinist system was that the trust (or enterprise) not the state became the basic financial and operating unit responsible for the payment of wages, with its own separate system of accounts.

The decision to reintroduce monetary relations between state-owned enterprises appears at first sight to extend the use of monetary relations beyond the degree necessary for the re-establishment of a stable-link between the town and the country and has significant repercussions for the operation of Soviet-type economies to this day.

In theory, in a socialist economy there is no *a priori* reason why the transfer of inputs and outputs between state enterprises should be accompanied by monetary transactions, and further why workers should receive payment from the enterprise instead of directly from the state budget. Indeed, Preobrazhensky had envisaged a socialist economy operating without monetary relations between nationalised industries. In 1921 he argued: 'All the nationalised undertakings will have a common counting house and will have no need for reciprocal purchases and sales (just like the single enterprise of a wealthy owner)' (Bukharin and Preobrazhensky, 1970, pp. 390–1). It was admitted that the problem was more complex in the case of a socialist sector in a mixed economy where money would have to survive for the part it would play in relations between the socialist and the private sector. Lenin, under the conviction that he was reintroducing State Capitalism may have felt this to be sufficient reason for the widespread use of monetary relations. There were, however, additional economic reasons for using monetary relations between state-owned enterprises. From the microeconomic point of view it was difficult to determine costs of operation and identify inefficient plants and activities without an individual system of accounting at the enterprise level, while the unintentional subsidisation of loss-making enterprises had inflationary consequences. During War Communism this had been expressed by a tendency for revenues to be distributed within the enterprise and for losses only to be transferred to the state budget.

The principal economic tasks facing NEP were to generate a sufficiently large surplus for investment in general, to provide for the growth of industrial capacity, and to stimulate an agricultural surplus to feed the growing industrial population and to provide a source of industrial raw materials and export earnings. The major political problem was whether the increased role of market relations and private capital would imply a return to capitalism if the size of the private sector were to outstrip that of the state sector, or would eventually facilitate such a return by making industry dependent on private traders who controlled large sections of retail and wholesale trade, while in the countryside the regime would come to depend on the richer capitalist farmers (or kulaks) for a marketed surplus of agricultural products.

The related questions of how to achieve these economic objects, of the desirable tempo of industrialisation and of the political implications of the chosen policies formed the basis of the 1920s 'industrialisation debate'. The view of the right wing or the 'geneticists', whose principal theorist was Bukharin,[1] was that in the short run the Soviet Union's comparative advantage lay in agricultural development and that the development of a flourishing agricultural base was a necessary prerequisite for a policy of industrialisation. The corollary of this was the preservation of an alliance with the peasantry through the continuation of price policies that would encourage the peasant to market agricultural commodities in exchange for consumer goods and agricultural implements, tractors, fertiliser, and so on. This implied the expansion of peasant demand to stimulate industrial investment and that the structure of investment should to a considerable extent be determined by the structure of peasant demand and be concentrated in light industry. Although Bukharin himself described these policies in unheroic terms — 'creeping at a snail's pace' and 'riding into socialism on a peasant nag' — the slow rate of growth referred to the size of the socialist sector vis-à-vis the capitalist sector, rather than a slow rate of economic growth, and Bukharin himself argued that these policies would achieve a 'very rapid tempo of development' (Cohen, 1980, p. 183).

If Bukharin's analysis was correct the major objection to his policies would appear to be that to encourage the peasant to pursue his own interest and enrich himself would stimulate the development of capitalism in the village, would strengthen the power of the kulaks and the private traders who were hostile to the regime, and that the structure of investment determined by such a policy would prevent the development of heavy industry, leaving the Soviet Union vulnerable to outside attack. Finally, the policy implied that the terms of trade between the town and the village (ie, the price ratios for industrial and agricultural commodities) should be influenced in favour of the latter, which would threaten to alienate the very industrial proletariat in whose name the Bolshevik regime ruled.

The conviction that the continuation of NEP would lead to a restoration of capitalism and that the security of socialism could only be guaranteed by a rapid rate of industrialisation with investment concentrated in heavy industry was central to the arguments of the 'left opposition' and its principal theorists, Trotsky and Preobrazhensky. Their policy was intended to ensure that the socialist sector grew more rapidly than the private sector by increasing the numbers of the industrial proletariat and that concentrating investment in heavy industry would ensure a rapid growth of the country's industrial and defence potential. To finance this investment Preobrazhensky proposed the policy of 'primitive socialist accumulation' defined as 'accumulation in the hands of the state of material resources mainly or partly from sources lying outside the complex of state economy' (Preobrazhensky, 1965, p. 54).

In terms of the relationship between agriculture and industry, Preobrazhensky (1965 edition) proposed that this should be achieved by a price policy based on unequivalent exchange which would involve a transfer of surplus value from the peasantry to the state, to be brought about either by taxation or through the exercise of the state trading monopoly to effect the terms of trade against the peasant. This policy therefore involved cutting back rather than stimulating peasant demand, in order to squeeze a source of investment finance to meet state determined tasks in heavy industry, with the possibility of a long period between the outlay of resources for investment and any return in the form of commodities for consumption. The critical question became whether the peasant would produce for the market and in particular the state market under these circumstances, or whether he would withdraw basically into production for self-consumption or for unregulated market activity and therefore would coercion be required to extract an agricultural surplus?

The debate was resolved by political manoeuvering against the background of changing economic conditions that have recently been the subject of much detailed research which can only be briefly referred to here. In the latter half of the 1920s a series of low grain-marketings threatened the supply of staple foodstuffs to the town. By 1926 the volume of grain output had approximately achieved pre-war levels while the volume of grain marketed to the town was only just over half the pre-war level (Davies, 1969). One reason was that grain prices had been kept deliberately low and the peasant had responded by shifting production to meat and dairy products, while shortages of consumer goods further lowered the incentive to market what grain was available. While the right wing pressed for grain price increases, Stalin, in a speech in May 1928, quoted figures that indicated that a significant proportion of pre-war grain-marketings had come from landowners' estates and kulak farms, while the redistribution of land that had taken place since the Revolution had largely resulted in peasants producing for their own consumption (Dobb, 1966, p. 217). The significance of this conclusion for the problem of industrialisation was profound. Firstly, it questioned the whole basis of whether a sufficient gross volume of agricultural supplies to the town could be secured through market policies and implied that industrial development based on the latter would require further agricultural inequality and would involve considerable dependence on the kulak. Secondly, it appeared to confirm the Marxian view that the principal source of pre-war investment had been brought about by forced expropriation from the poorer peasant — the corollary of which was 'primitive socialist accumulation'. The further significance of this was that the equilibrium prices that were being urged by the right wing would have required more resources (particularly in view of the goods famine) to be diverted to meeting peasant demand, certainly at the expense of the town, and possibly at the expense of aggregate investment.

Stalin's solution in 1928 was the re-use of compulsory requisitioning to extract an agricultural surplus combined with a series of discriminatory measures against the richer peasant which further damaged any prospect of encouraging increased agricultural productivity through the use of money incentives. It was against this background that the policy of collectivisation of agriculture as a counterpart to centralised planning of industry evolved.

The Basic Outlines of the Stalinist Economic System

The broad principles of the Stalinist system of industrial development were to concentrate a significant volume of national income into investment with additional priorities to investment in industry and construction in general and into industrial producers' goods in particular.

This pattern of development was largely derived from Stalin's interpretation of Marx's model of expanded reproduction, which implies that continued economic growth requires that the output of producers' goods (Department I, or goods to be used in the process of further production) should grow faster than the output of consumption goods (Department II). This rule poses a number of problems of identification and measurement that make it difficult to translate into practical policy measures. It does not necessarily imply the primacy of industrial output over agricultural output as agricultural output can go to either department — it might however be used to justify a preference for industrial crops over food crops. Similarly, most industrial statistics group commodities by their sector of production or origin, not sector of destination (eg, both coal and refrigerators could be either means of production or means of consumption).

Fallenbuchl (1970) has provided an assessment of how Soviet planners in the Stalin era attempted to translate this aspect of Marxian theory into practice; three operating principles have resulted:

1. A high rate of investment expressed in terms of the concentration of a significant proportion of social product into investment is considered necessary as a means to rapid growth.
2. Investment should be concentrated in material production —

ie, it should lead directly to the production of physical products (productive investment) rather than infrastructure, services, and so on (non-productive investment).

3. Priority of investment in industry should be concentrated on the means of production (ie, sectors producing producers' goods). It is Soviet (and CMEA) practice to divide industrial output between the production of producers' goods (Group A) and the production of consumers' goods (Group B).

Fallenbuchl (1970) shows that Soviet planners have accepted as a rule of thumb that Department I for the economy as a whole will grow faster than Department II, if sufficient investment resources are concentrated in industry and the output of Group A grows faster than Group B. Nove (1980, p. 336) has indicated that Group A should not be directly equated with heavy industry or Group B with light industry. Stalin however rejected a 'calico' industrialisation and imposed his own priority for heavy industry and his own preference for the production of machine-building ferrous and non-ferrous metallurgy, fuel and power.

As a result, Soviet investment priorities in the first five-year plan indicate an extreme interpretation of this model of the priority of Department I, both in terms of the size of national income devoted to investment, and in the structure of industrial output deemed to constitute 'Group A'. Ellman (1975) has calculated using 1928 prices (which may overestimate the share of industrial output and its growth rate) that between 1928 and 1932 net material product in the USSR grew by 15 billion roubles (from 24.1 to 40.1 billion) while by sector of use investment grew by 14 billion roubles (from 3.7 to 17.7 billion). Furthermore, primarily as a result of the fall of production of meat, milk and eggs, the value of consumption in 1932 was below the level of 1929. Measured by sector of origin, net industrial output expanded from 7.6 billion roubles to 17.8 billion and construction from 1.9 to 6.0 billion roubles. Soviet official statistics indicate a growth of industrial production of 19.5 per cent per annum and of 28.5 per cent and 11.7 per cent for Group A and Group B respectively, while Group A received 86 per cent of industrial investment. Although the gap between the growth of industrial producers' goods and industrial consumers' goods narrowed in the second five-year plan, the three principles referred to by Fallenbuchl have not actually been broken in Soviet practice in any plan period to date. Fallenbuchl (1970) also indicates Stalin's clear preference for developing the metallurgical and machine tool industries, which in addition to requiring a rapid growth of iron and steel output also required the development of coal, coke, power and fuels as basic inputs. Consequently the growth rate of these sectors was drastically increased over the levels intended in the original plan variants.

This virtual personal preference for a specific industrial structure was of considerable importance when the model was transferred to East European countries with far less favourable resource endowments than the USSR.

This industrial policy was accompanied by the collectivisation of agriculture, which was intended both to enlarge the socialist sector of agriculture and to increase the flow of resources from agriculture to industry, the latter to be achieved by increases in productivity, which were expected to result from the higher degree of specialisation and mechanisation that would be brought about by large-scale farming, and more importantly by the production and exchange policies that could be imposed on collective farms by Communist Party management. It is important to distinguish here between the gross flow of resources from agriculture to industry (total sales of agricultural outputs to industry) and the net flow (the value of total sales of agricultural outputs to industry minus agricultural purchases of industrial commodities) (Millar, 1970). Preobrazhensky's policy of primitive socialist accumulation, together with his desire to maintain a high level of wages to industrial workers, required that the agricultural sector should make a *net* contribution to both the state sector and the industrial population as a whole (Davies, 1980, p. 34). Preobrazhensky (1965, pp. 110–11) did not however anticipate that this should be achieved by collectivisation but by price policy. In addition the policy of industrialisation which required a flow of labour from agriculture to industry would have been impossible without a substantial gross flow of resources emanating from the agricultural sector, principally in the form of basic foodstuffs to feed the growing industrial population, but also in the form of raw materials where such resources could not be obtained from imports and in the form of a contribution to the balance of payments if foreign lending was not forthcoming.

To what extent were the industrial policies pursued by Stalin in the first five-year plan, and in particular, the policy of collectivisation, determined by economic or political 'necessity' under the conditions prevailing in the USSR and what lessons do they offer either for economic development per se or for an economy developing on socialist lines? Nove (1964) argues that, given the internal and external opposition to the regime, rapid industrialisation was necessary to strengthen the size and position of the industrial labour force and to develop and maintain defence industries. This, it is argued, required a substantial volume of investment in heavy industry which in turn required an increased flow of agricultural commodities which would not have been generated by voluntary policies. Nove also argues that certain aspects of the policy, notably over-rapid industrialisation and over-rapid collectivisation combined with the policy of the elimination of the kulak, which led to the wholesale slaughter of livestock and destruction of agricultural implements with disastrous effects on agricultural productivity, could not be considered necessary but must be attributed to errors of judgement and hostility towards the more efficient farmer. In its broad outlines, however, Nove considers that rapid industrialisation and collectivisation were necessary under the prevailing conditions.

The contribution of agriculture and collectivisation to economic development in the first five-year plan has been the subject of renewed research and debate by Western scholars since the publication of new Soviet data in the late 1960s.

J. R. Millar (1974) has attempted to evaluate what he terms agriculture's net contribution to industrialisation by measuring the flow of resources from agriculture to industry minus the flow of resources from industry to agriculture (the *net* agricultural surplus). His principal argument can be summarised by saying that if 1928 price-weights (ie, those prevailing in the last year before collectivisation) are used to measure inter-sector flows, the agricultural sector was actually a net importer of resources from industry in each year from 1928 to 1932, and that the deficit actually increased following collectivisation. His explanation for this is that price controls and food shortages caused free market prices for agricultural products to rise so sharply that the overall terms of trade actually moved in favour of the village, and as a result the peasant was able to pass some of the costs of collectivisation back on to the industrial worker, and that industrial inputs (tractors, etc) had to be provided to collective farms in the form of investment to compensate for the destruction of livestock primarily caused by opposition to collectivisation itself. Millar concludes therefore that collectivisation was an 'unmitigated economic policy disaster' (1974, p. 764) and more significantly that 'a continuation of the New Economic Policy would have permitted at least as rapid a rate of industrialisation, with less cost to the urban as well as the rural population' (1974, p. 766).

Ellman (1975) argues that the balance of trade between agriculture and industry (the net agricultural surplus) is not an appropriate measure of agriculture's contribution to either net investment in the economy, or to the resources available for industrialisation. The former should include the value of net investment in the agricultural sector (which Millar considers to be negative) and the latter should include the value of industrial producers' goods delivered to agriculture (ie, these should not be deducted from agriculture's contribution towards a *policy* of industrialisation as they actually stimulate industrial output).

Furthermore, Ellman (1975) argues that the 1928 agricultural/industrial price ratios were not representative of NEP as a whole because a price policy that was unfavourable to the village was being pursued in that year. Consequently, he argues, the use of 1928 prices considerably underestimates the real value of agricultural flows to industry and also underestimates the size of primitive socialist accumulation arising out of the agricultural sector as that policy was already being pursued before the collectivisation drive.

Ellman and Millar agree that the act of collectivisation did not yield an increase in the *net* agricultural surplus to help to finance investment in the first five-year plan. Millar considers that collectivisation had no economic rationale, while Ellman indicates that the volume of investment undertaken in the first five-year plan required inputs of both labour and resources from agriculture, which he argues would not have been forthcoming without collectivisation. He therefore considers that collectivisation contributed to industrialisation by forcing labour to move out of agriculture and by providing a *gross* flow of resources (basic foodstuffs, exports and import substitutes) to industry and the urban sector.

The finance for investment in the first five-year plan was provided in part by turnover tax (levied mainly on agricultural commodities consumed by the industrial population) and to a considerable extent by inflation which contributed to a drop in the real wages of the urban population. Thus, as Millar indicates, some of the burden of collectivisation was borne by the urban population, which effectively meant that the population as a whole financed investment, which, as Ellman argues, is entirely consistent with Preobrazhensky's concept of primitive socialist accumulation. Whether a sufficiently large *gross* agricultural surplus to permit rapid industrialisation could have been achieved without collectivisation is still open to debate, but under Soviet circumstances this would have implied considerable political dangers by strengthening the role of the kulak and the private trader.

On balance it appears that Millar's calculations do not prove that socialist accumulation (defined either as the extraction of a surplus from the population *as a whole*, or even from the agricultural population) did not take place during the first five-year plan. He does however demonstrate that collectivisation did not either increase the size of the surplus or generate the finance for investment and throws serious doubt on the validity of collectivisation both under Soviet conditions and as a model of development in general. Although agriculture made a considerable gross contribution to industrialisation, the question remains whether under different conditions in other countries such a contribution might be achieved by measures that would entail a lower human and economic cost.

The role of money in the economy was a further subject of debate during the first five-year plan, while the method of financing industrial investment was subject to repeated modification between 1927 and 1939. During NEP the banking system had played a key role in raising and distributing finance and had succeeded in concentrating a substantial proportion of industrial investment (74 per cent) in producers' goods (Fallenbuchl, 1970, p. 364). The advent of central planning however was to put increased stress on the system in regard to both the source of funds which were to be 2–3 times larger than those raised under NEP, and the direction of investment which was to be far more rigidly determined by central criteria. In theory it would appear that this problem could be best solved by a greater role for the state budget in directly raising revenue through taxation and disbursing them according to plan priorities.

Davies argues that at the end of the 1920s no clear economic model existed in the minds of Party theorists who had expected NEP to last for some considerable time and be eventually replaced by a moneyless economy.[2] As the collapse of NEP became inevitable, they therefore anticipated a rapid transition to a moneyless centrally-planned economy with the abolition of monetary accounts between enterprises and the abolition of private trade and its replacement by direct product exchange between industry and a collectivised socialist agriculture.

The concept of primitive socialist accumulation did not imply that investment would be entirely raised by unequivalent exchange with the agricultural sector, but that reduced costs of industrial operation brought about by productivity improvements and large-scale production would also make a contribution. In the initial stages of planning it was essential that the centre gained information about, and control over, the operation of enterprises. Podolski (1973, p. 24–8) argues that the failure to meet anticipated productivity levels in 1928–9 and the resulting inflation led to the decision in December 1929 to make the individual enterprise the basic production unit, with its own system of accounts and responsibility for the fulfilment of plan tasks, but monetary criteria were to remain subservient to physical instructions.

The critical questions concerned the financial relationship between the enterprise and the state and in particular whether enterprise income should be paid directly to the state budget, be lent to financial intermediaries for reinvestment or be retained by the enterprises themselves. The latter course would imply a loss of central control over investment and risk the dissipation of productivity gains in the form of increased wages rather than accumulation, while the experience of War Communism indicated that moneyless inter-enterprise transfers combined with high investment rates and centrally paid wages was inherently inflationary. The logical response was that while money was retained enterprise profits should be paid into the state budget, while controls were maintained over wage payments by a system of separate enterprise accounts.

The failure of collectivisation to stimulate a sufficient marketed surplus to prevent famine in 1932 and the resulting flight from the towns led to the reinstatement of private peasant plots and the legalisation of private grain-trading. This logically required the development of a market for consumer goods, the perpetuation of a wage payment system and monetary relations between enterprises whose control was continually refined in the period up to 1939.

The implications of the decision to continue the role of separate enterprises, whose existence in NEP had largely rested on Lenin's conception that he was reintroducing State Capitalism, was of considerable significance for the structure and operation of a socialist economy and for the centrally-planned economies of Eastern Europe in particular. A considerable body of western analysis of the operation of centrally-planned economies cites as major weaknesses of the system, those areas where the interests of enterprises or enterprise management diverge from those of the economy as a whole, while much domestic discussion of reform of the East European economies has concentrated on the division of decision-making between planning agencies and enterprises or groups of amalgamated enterprises, and on improving the harmony of interests between central authorities and enterprises. In addition more far reaching concepts of reform involving workers' control or participation in decision-making view the continued existence of separate enterprises as desirable ends in themselves with humanitarian consequences.

Notes

1. For an account of Bukharin's economic theories see Cohen (1980), especially Chapter 6.
2. Lewin (1975) Chapter 5 makes a similar suggestion. This is now the subject of detailed research by Professor R. W. Davies, to whom I owe this point.

2 THE ORIGINS OF THE STALINIST MODEL IN EASTERN EUROPE

Economic Development in Eastern Europe in the Inter-War Periods

The East European nations that subsequently adopted the Stalinist model of economic development (with the exception of Czechoslovakia), were poor, predominantly agrarian nations in comparison with Western Europe. The major indicators of economic development at the eve of the Second World War are shown in Table 2.1, which indicates that the nations may be divided into three distinct groups. The least developed were the Balkan states, Bulgaria and Romania, with a national income (in 1948 US dollars) of less than 70 dollars per head; Poland and Hungary who were peripheral to central European development had a per capita income of over 100 dollars; and Czechoslovakia, the most industrialised nation, had a per capita income of 176 dollars (compared with 378 dollars and 236 dollars per head in the UK and France respectively) (United Nations, 1949, p. 235).

The levels of income per head corresponded to the levels of industrialisation in the individual countries. Under 10 per cent of the labour force in Bulgaria and Romania were employed in industry and mining, and 80 per cent of the population depended on agriculture as a means of support, compared with 33 per cent in Czechoslovakia.

Inequality of land holdings in the area as a whole was reflected by the fact that 70 per cent of land holdings, accounting for 20 per cent of the total acreage, were subsistence farms of under 5 hectares, while less than 1 per cent of agricultural holdings accounted for 40 per cent of the acreage. Inequality of land holdings was most extreme in Hungary, where 85 per cent of holdings were under 5 hectares while the larger estates extended to over 3,000 hectares. The degree of mechanisation of agriculture was greatest in Hungary and Czechoslovakia (where agricultural inequality was greatest) while the proportion of agricultural land

Table 2.1: General Indicators of Economic Development in Inter-War Eastern Europe (Inter-War Boundaries)

	Bulgaria	*Romania*	*Poland*	*Hungary*	*Czechoslovakia*
Per capita income (1948 US dollars)	68	60–70	104	112	176
% population dependent on agriculture	73	72	60	52	33
% labour force in agriculture	80	80	65	51	26–37[a]
% population in labour force	57	58	47	46	44
Estimated disguised unemployment as % of agricultural labour force	28	20	24	18	13
Total population (millions)	6.2	19.2	34.5	8.9	15.2
Land distribution					
% holdings under 5 hectares	63	75	65	85	70
% land in holdings under 5 hectares	30	28	15	19	16
'Relative Inequality'; 5 = most equal	5	4	3	1	2
'Relative' Mechanisation of agriculture; 5 = least mechanised	3	4	5	1	2
% holdings over 100 hectares	under 1%	0.4	0.6	0.5	0.5
% land over 100 hectares	under 1%	28	45	43	41
Labour production (1938 US dollars) (per employee)					
Large-scale industry	320	430	500	430	550
All industry	300	290	400	340	450
Agriculture	110	80	130	150	200
Output per hectare	45	27	52	43	67

Note: a. See p. 20.

Sources: All figures derived directly from, or estimated from data in Spulber (1957) except for: per capita income (United Nations, 1949); disguised unemployment Rosenstein-Rodan (1943).

devoted to subsistence farming was highest in Bulgaria and Romania and lowest in Czechoslovakia.

There is some prima facie evidence from these highly aggregated figures to suggest that the model of economic development espoused by Stalin, involving the forcible transfer of labour from agriculture to industry and a high rate of industrial investment, would succeed in generating economic growth in Bulgaria and Romania and possibly Hungary and Poland, but would be of limited value in Czechoslovakia.

Some words of caution are required. Rosenstein-Rodan's estimates of hidden unemployment, shown in Table 2.1, have been criticised on the grounds that they overestimate the amount of labour available for industrial employment by underestimating seasonal peak demands for agricultural labour and by including 'assisting family members' in the agricultural labour force. The impact of assisting family members can be seen in rows 2, 3 and 4 of Table 2.1 by the higher proportion of the agricultural population registered in the labour force in the more agrarian countries, and by the higher ratio of the percentage of the labour force employed in agriculture to the percentage of the total population dependent on agriculture in the more agrarian countries. In Czechoslovakia the percentage of the labour force employed in agriculture falls from 37 per cent to 26 per cent when 'assisting family members' are excluded.

Jorgensen (1971) has suggested that the effect of seasonality may turn labour surpluses into temporary labour shortages, while more recent studies suggest that the proportion of labour that could be moved out of agriculture without a decline in output rarely exceeds 5 per cent (Elkan, 1973, pp. 70–1).

In Hungary, for example, the larger estates concentrated on grain production and did not yield a high value of output per hectare. The smaller farms which concentrated on higher value, labour-intensive crops, yielded approximately double the value of output per hectare, and also marketed 40 per cent more output per hectare in value terms than the larger estates (Dorner, 1972, p. 125). This indicates that agricultural labour did have a positive marginal product, that small farms did produce for the market, and that improvements in income might have been obtained by a policy of land redistribution and mechanisation of agriculture combined with a market system.

Inter-War Industrial Development Policies and Foreign Trade

Spulber (1957) estimates that the industrial labour force employed in medium and large-scale industry in the area (excluding the territory that subsequently became the GDR) totalled about 3 million workers, of whom 1.2 million were located in Czechoslovakia. Industrial employment in Czechoslovakia, Hungary, Poland and Romania was largely concentrated in large plants employing more than 500 workers. Output per worker was highest in (capital-intensive) large-scale industry and the ratio of output per worker in industry to output per worker in agriculture was highest in the least developed country (Romania) and lowest in the most developed (Czechoslovakia). In Poland and Bulgaria the industrial workforce was largely employed in the production of consumer goods, but elsewhere was equally divided between the production of producers' goods and consumer goods.

The world recession and the small size of the domestic market meant that large-scale production and the development of new industries had to be fostered by a number of protectionist measures. The engineering and food-processing industries in Hungary, for example, had developed before the First War to meet the demand of the entire area embraced by the Habsburg Monarchy. The textile, chemical and electrical industries developed in the inter-war period against a background of import quotas and exchange controls and did not develop a competitive market mentality, but sought to maintain high profit levels through high prices rather than through improvements in productivity and cost reductions (Kemeny, 1952, pp. 1–5).

In Romania the impact of protectionist policies enacted between 1929 and 1933, associated with Manoilescu (Montias, 1967, pp. 195–6), bore strong similarities to the policies of accumulation advocated by Preobrazhensky and List, but were conducted in a state-directed, but essentially market framework. State intervention favoured the development of industry at the expense of agriculture, on the grounds that this would result in a higher potential growth of output per man. Import quotas were largely imposed on consumer goods whose share of total imports fell from 46 per cent to 21 per cent from 1929 to 1938, while the share of machinery and equipment rose from 44 per cent to 56 per cent and raw materials from 10 per cent to 23 per cent. (Total imports declined by about 10 per cent in real terms.) Industrial output grew by about 33 per cent with the output of producers' goods (44 per cent) growing faster than consumer goods (23 per cent). The monopolisation and cartelisation of industry influenced industrial prices more favourably than agricultural prices and Romanian industry was largely uncompetitive in world markets in both price and quality and operated with considerable excess capacity (reaching 50–70 per cent in the consumption oriented industries, including food-processing and brewing). The industrial policies of the early 1930s operated by restricting rather than expanding consumption which grew more slowly than the total population as a result of which real income, and in particular peasant incomes declined (Lupu, 1969).

The Bukharinist view was put by Madgeauru (1930), the Minister of Finance in the National Peasant Party, who attacked 'the chimera of forced industrialisation with economic self-sufficiency in view' (p. 6), and argued that priority to industry and tariff barriers against agricultural machinery were responsible for the low level of mechanisation and decline in Romanian agriculture. He proposed to improve the prices of agricultural goods relative to industrial goods by removing tariffs and quotas, prohibiting cartelisation and giving priority investment to agriculture. This would stimulate peasant income which, in turn, would stimulate the demand for light industry and, Madgeauru (1930, p. 62) argued, would create 'new groups of consumers in the agrarian countries of eastern and southern Europe', which would lead Europe out of depression.

Similarly, in Hungary protectionist policies encouraged industrial development at the expense of the consumer. Hungary's restricted domestic resource base meant that it was largely dependent on imported raw materials which constituted 44 per cent of all imports between 1933 and 1937, while the lack of competiveness of Hungarian industry meant that 60 per cent of its exports over the same period were composed of unprocessed foodstuffs which reduced the supply of these commodities to the domestic population and increased their real contribution to industrial investment (Kemeny, 1952, pp. 3–5).

In the second half of the 1930s the less-developed economies of South East Europe became increasingly dependent on the German economy from whom they imported manufactured goods, mainly in exchange for bulk sales of raw materials and foodstuffs. As a result, roughly one-third of Romanian and Bulgarian exports were composed of foodstuffs and two-thirds raw materials and semi-processed goods (Zauberman, 1964, p. 272). Poland, who was more successful in resisting integration with the German economy, traded more extensively with Britain, but still had a similar export structure, with a greater degree of importance for manufactures (6 per cent). In Czechoslovakia manufactured goods comprised 72 per cent of total exports in 1938, despite the protectionist policies of her neighbours and the integration of the Balkan states into the German economy, which cost Czechoslovakia roughly half her export trade.

The demands of the German war economy for armaments stimulated public investment programmes in heavy industry in the axis powers and occupied lands, and strengthened the role of the state as an initiator of economic development, which influenced the attitudes of many communist and non-communist leaders alike in the post-war era.

Economic Policy, 1945–8

In the immediate post-war period power was exercised by National Popular Front Parties involving coalitions of principally Communist and Socialist Parties but also extending to other minority parties representing smallholders', workers' and even landowners' interests. A parliamentary system survived in Czechoslovakia and Hungary until 1948 and the monarchy survived in Romania until 1947.

Communist and Socialist leaders professed open commitment to a mixed economic system in which public and private ownership would coexist with a dirigiste system of planning based on monetary levers rather than physical instructions involving the nationalisation of the central bank and monetary and credit reforms.

Currency reforms were required in most countries to establish a stable currency in the presence of massive inflationary pressures generated by excess monetary emission, the collapse of industrial and agricultural output and reparations imposed on the axis powers. Credit reforms were also required to direct investment towards centrally-determined objectives and to repair war damage.

There was also fairly general agreement between socialists and communists that government policy should stimulate industrialisation to provide employment for the 'surplus agriculture population' and to stimulate economic growth, although there was considerable disagreement on whether the industrialisation policy should be concentrated towards producers' goods or consumer goods and on the importance of agricultural development.

As a simplification, the system proposed by communist leaders may be compared with Lenin's (and Hilferding's) concept of State Capitalism which proposed to stimulate the supply of capital to a mainly private, but highly concentrated industry through the control of the banking system, while the system proposed by the socialists resembled the policy proposals of the socialist movements in Western Europe.

The social-democratic view in Hungary has been summarised by Kemeny (Under-Secretary for Finance in the Hungarian coalition government from 1945–8) (1952, pp. 6–16). Only such branches of industry as were 'monopolistic in character' were to be nationalised, involving 'almost the whole of mining, approximately 35% of manufactures, 80% of banking and the whole of rail and river communications', together comprising less than a third of total production. Controls over credit, investment, wages and prices, combined with direct allocation of certain raw materials, would achieve government goals without further nationalisation.

In Czechoslovakia the Communist Party Secretary, Gottwald, proposed in 1945 that the policy of nationalisation should involve fixed limits between the state and the private sector and that private initiative would play a decisive role in the economy. To prevent uncertainty, nationalisation would have to be undertaken in a single move. In practice, the criterion of nationalisation in Czechoslovakia was plant size and by the end of 1947 nearly two-thirds of all employees worked in nationalised undertakings (Spulber, 1957, pp. 47–50).

Hilary Minc (Poland) proposed in 1946 that the economy would compose elements of private, state and co-operative property and that the private sector would embrace 40 per cent of the industrial labour force. In practice, by the end of 1947 only 10.5 per cent of the labour force was privately employed (Spulber, 1957, pp. 59–61). The degree of nationalisation was far lower in the Balkans extending to only 6 per cent of the output of Bulgarian industry at the end of 1947 and largely involving the joint Soviet-Romanian companies (Sovroms) in Romania.

In many cases, however, Communist Parties attempted to

introduce a far greater number of central controls over state and private industry. In Romania Georghiu-Dej proposed to establish a Ministry of Industry and Commerce that would bring the various commissariats for prices, foreign trade, material supplies, and so on, under the control of a single agency and turn it into an effective organ for control of the economy, which would extend to drawing up production programmes, controlling foreign trade, and procuring and distributing raw materials. Industry was to be administered by a number of central industrial offices which were to function as limited companies headed by an administrative council composed of private owners and representatives of the Ministry of Industry and Commerce (Academia, 1964, pp. 38–42).

Fourteen industrial offices, embracing 750 enterprises and 80 per cent of industrial capacity, were established in December 1947 despite opposition from the non-communist parties. Their powers were quite extensive and included drawing up proposals for the organisation of production; the distribution of fuel and raw materials by branches and enterprises; the procurement and supply of raw materials; the distribution of finished products; controls over the quantity and assortment of imports and exports; measures for the rationalisation of production and proposals for new investment. The operation of individual enterprises was to be supervised by workers' committees reporting back to the Ministry.

A major area of controversy between Communists and Socialists in the coalition governments concerned the priority to be given to consumption and investment and the relationship between agriculture and industry. The socialist parties placed greater emphasis on the restoration of pre-war consumption levels and thereafter on 'balanced growth'. Spulber (1957, p. 287) shows that in each country of the bloc the initial reconstruction plans drawn up by the coalition governments (roughly covering the period from 1947 to 1949) placed greater emphasis on agricultural development than even the initial five-year plans drawn up by the domestic communist parties and the subsequent revised versions drawn up at Stalin's behest after the outbreak of the Korean War.

Although these differences can partly be attributed to the overwhelming need to restore food output to pre-war levels, before further developments could be contemplated, they also reflect a clear difference of outlook. In Hungary, the Social-Democratic Party drew up their own plan proposals which proposed a far lower rate of investment and a more rapid growth of light industry than was contained in either the Communist proposals or the finally approved reconstruction plan that lay nearer to Communist than Socialist proposals. The basis of the Social-Democratic proposals was that the relative scarcity of capital, raw materials (including coal and iron-ore) and abundance of labour pointed to the development of labour-intensive industries such as textiles, in preference to heavy industry, and that a flow of consumer goods would stimulate labour productivity (Kemeny, 1952, pp. 23–5). A further argument was that agriculture should develop along labour-intensive lines, similar to the Danish pattern, concentrating on the production of vegetables, livestock and on horticulture (Balassa, 1959, p. 245).

Drewnowski argues that disputes over the pattern of industrialisation were an area of bitter disagreement between the Socialist and Communist Parties in Poland. The Socialist Party dominated the Central Planning Office which was responsible for preparing economic plans, while the Communist-dominated Economic Committee of the Council of Ministers and Ministry of Industry were responsible for the implementation of economic policy. The initial reconstruction plan (for 1947 to 1949) gave priority to the rapid increase of consumption; the annual draft for 1948 was attacked, in February 1948, by Communist Party members of the economic administration on the grounds that it was based on western techniques of national income accounting, it did not contain proposals for ensuring its fulfilment, and that it placed too much emphasis on consumption and did not give sufficient priority to production. The Central Planning Office was reorganised a month later and replaced in April 1949 by the State Commission for Economic Planning, a body similar to the Soviet Gosplan (Drewnowski, 1979).

The long-term implications of the more consumption-oriented policies of the Socialist Parties, and of the Hungarian Social Democrats in particular, that the less-industrialised nations should concentrate on the production of labour intensive light industrial and agricultual commodities were that they should import machinery and equipment and engineering products from Western Europe and Czechoslovakia (who would have been ideally placed to fill the vacuum left by the collapse of the German economy) and were supported by proposals for the development of customs unions embracing groupings of East European nations.

These proposals were unpopular with the nationalist inclined members of the governments of the less-developed countries who had been forced, by steadily deteriorating terms of trade in the 1930s, into a position of economic and political dependence on Germany and had entered into bilateral deals in which they supplied raw materials in exchange for machinery and equipment.

In Romania, Georghiu-Dej (Academia, 1964, pp. 29–31) attacked proposals which he argued placed emphasis on the concept of Romania as a 'predominantly agrarian nation' at the first post-war conference of the Romanian Workers (Communist) Party in October 1945 as 'representing the interests of large landowners and capitalist classes'. Bourgeois intellectuals who had proposed industrialisation as a means to national self-determination were described as patriots; their opponents who emphasised agrarian development were denigrated. The conference confirmed support for a policy of industrialisation based on priority to heavy industry (regardless of the domestic raw material base) not purely as being 'the correct Marxist-Leninist course' but as the most appropriate method of using Romania's human and natural resources on grounds similar to those advocated by Manoilescu.

The integration proposals were also unwelcome to Stalin who proved unwilling to allow either the degree of trade this would involve with Western Europe or the degree of co-operation and integration this would involve between the East European states themselves, which in particular could have led to the emergence of a strong Czechoslovak economy.

The economies of the former Axis powers were further weakened by the need to pay reparations to the USSR and the Soviet acquisition of former German holdings in East European countries, which formed the Soviet contribution to the joint companies whose production was then shared between the USSR and the 'host' country. In this way production patterns which had formerly been integrated with the German economy were forcibly integrated with the Soviet economy.

Land Reform

The most urgent priority facing the coalition governments in the immediate post-war period was land reform. In Poland this involved the colonisation of land acquired in the former German territories and in Czechoslovakia the redistribution of land formerly belonging to German and Hungarian nationals. In Bulgaria, Romania and Hungary land reform involved the break-up of large estates and their redistribution to landless and poorer peasants, resulting in the creation of a large number of subsistence (or even below subsistence) farms (Spulber, 1957, pp. 235–50).

The extent of land redistribution roughly corresponded to the degree of pre-war inequity; in Bulgaria only 2 per cent of agricultural land was affected, and about 7 per cent in Romania, where 1.5 million hectares were confiscated and the majority redistributed to 0.4 million landless peasants and half a million peasants with very small holdings. A more substantial redistribution took place in Hungary where an upper limit on land holdings of just over 100 acres was imposed and 1,500 large estates (involving 30 per cent of the cultivated area) were redistributed to 370,000 landless peasants and 270,000 poor peasants (Academia, 1964, pp. 25–6).

These measures were supported by Communist members of the governments who attempted to allay the fears of the peasantry that a policy of collectivisation might be forthcoming. Rakosi in 1945 specifically denied that Communists wanted to establish collective farms and argued that they really wished to encourage small prospering farms; Gero in 1947 argued that collective farms were consistent with Russian traditions but alien to the history and culture of the Hungarian people (Balassa, 1959, p. 246).

Spulber (1957, p. 246–7) argues that the Communists were partly forced to take up this position as a result of land policies in the 1930s which meant that peasants effectively owned the land they cultivated. Any attempt to nationalise, but not redistribute, land, would be interpreted not so much as an act against the landlord but against the peasant himself. Spulber also proposes that the Communist Parties (with the encouragement of Soviet analysts) supported the progressive redistribution of land with the intention of creating uneconomic units which would lead to the collapse of small-scale private farming.

This hypothesis is plausible but difficult to prove due to the paucity of data. Although it had been initially intended in Hungary that small-scale private farms would be encouraged by price incentives to produce for the market and to make deliveries to central authorities, Kemeny (1952, pp. 38–40) shows that this policy was ineffective as early as 1946.

Measures of compulsion and coercion, not unlike (but less severe than) those used under War Communism, were employed in Hungary to ensure agricultural provision for the urban population before collectivisation and before the final communist take-over of power. Plans expressing sowing targets and average yields were drawn up for major crops, together with output targets for milk, meat, poultry, eggs, and so on. A significant proportion of farm output had to be delivered to the state at fixed prices. Kemeny argues that deliveries were set sufficiently high to require efficient working for their fulfilment while prices were fixed at a low level resulting in a flourishing black market, which in turn encouraged peasant evasion of deliveries. As a result, he argues (p. 38), 'flying squads were organised to apply harsh methods in repressing peasants' attempts to slip through the meshes of the regulations requiring the delivery of grains, fat and other produce.'

Whether the promotion of small-scale agriculture was a deliberate tactic to ensure its eventual collapse or not, the official explanation for the socialist transformation of agriculture involving dekulakisation and setting up co-operative and collective farms in Romania in March 1949 was given as 'small scale peasant agriculture would not permit extended reproduction and made the utilisation of modern machinery and agricultural techniques impossible' (Academia, 1964, p. 71), while similar justifications were given elsewhere.

Stalinisation, 1948–53

The principal features of the Stalinist method of industrialisation, involving the wholesale nationalisation of industry, the establishment of a planning bureaucracy issuing physical instructions to enterprises under strict party control and the institution of a state monopoly of foreign trade were not implemented immediately after the communist take-over of power in 1948. For example, Kosta (Brus, 1980) argues that in Czechoslovakia this happened gradually after 1948 and under considerable Soviet pressure.

In agriculture collectivisation and dekulakisation through confiscatory tax measures proceeded slowly. Initial moves were made towards loose forms of collective farms, which frequently resembled pure 'tillage associations' with the crop distributed according to land holdings, but were followed by a move towards the Soviet-type 'artel' in which land, livestock and implements became common property with farm income distributed on the principle of work contributions. In Czechoslovakia, for example, the loosest form of co-operative farm accounted for 96.1 per cent of socialised agriculture in 1949, 45.7 per cent in 1950, 15.7 per cent in 1952 and was abolished in 1954 (Kirschbaum, 1980, p. 61).

Peasant hostility and problems of local administration were major causes of slow progress to collectivisation in each of the East European countries. At the death of Stalin only Bulgaria had collectivised more than 50 per cent of its agricultural land, while in Romania and Poland the proportions were estimated to be as low as 6 to 7 per cent, 21 per cent in Hungary and 34 per cent in Czechoslovakia. The importance of state farms (on which peasants are purely employees receiving wages paid on a piece-rate basis), which were established on expropriated lands, accounted for about 10 to 14 per cent of arable land in all countries except Bulgaria (2 per cent). In 1955 80 per cent of agricultural land in Poland remained in private hands, 50 per cent in Czechoslovakia and 60 per cent in Hungary (United States, 1955, p. 112).

Enforced collectivisation was resumed everywhere in the bloc, except Poland, in 1957. By 1966 state and collective farms accounted for 90 per cent of agricultural land in each country except Poland where the proportion remained at 14 per cent (Marczewski, 1974, p. 18).

The nationalisation of manufacturing industry was more immediate throughout the bloc. In Hungary, a decree issued in March 1948 nationalised all manufacturing enterprises employing more than 100 workers (embracing 84 per cent of industrial workers) and was extended to cover remaining enterprises with more than ten workers (99 per cent of industrial workers) in December 1949 (Balassa, 1959, p. 29). In Czechoslovakia the private sector embraced only 3.6 per cent of manufacturing employment at the end of 1948 (Spulber, 1957, p. 50). In Romania over 85 per cent of manufacturing industry was nationalised by the end of 1948 (Academia, 1964, pp. 53–5).

A small handicraft sector coexisted uneasily with the state sector and either fell under the tutelage of local administrations or remained in private hands, but subject to stringent controls and denied access to priority raw materials. In Hungary employment in the handicraft industries declined from 273,000 in 1948 to 42,000 in 1953 (Balassa, 1959, p. 33).

The Korean War and Priority to Heavy Industry

The precise circumstances surrounding what may be the most significant event in the post-war economic history of Eastern Europe remain something of a mystery. Kaser (1967, p. 49) reports that a meeting of Party leaders took place at Hollohaza, Hungary in November 1950 at which Stalin instructed each country to revise its targets for heavy industry in the 1951 to 1955 five-year plan substantially upwards in order to meet the requirements of the Korean War.

Czech economists (Selucky, 1972, p. 25) also attribute the adoption of the full system of Stalinist planning to the outbreak of the Korean War and Kosta (Brus, 1980) reports that the CMEA played a critical role in this process. Ausch (1972, p. 44) proposes that the general economic policies of the CMEA countries were developed 'between 1949 and 1951 influenced partly . . . by a preparation made necessary by the expectation of an imminent new world war.'

There were two separate strands to this process. The first involved the attempt to build up heavy industry, including the production of iron and steel and engineering products in the bloc as quickly as possible and the second involved the attempt to pursue these policies on an autarkic basis — ie, attempting to build up these industries in each country of the bloc without regard to the prevailing conditions (especially the supply of raw materials) in each individual country. Ausch (1972, p. 44) argues that the second aspect of the policy was in part determined by the attitudes of the East European leaders themselves. Hungary, for example, was unwilling to participate in the expansion of Polish coal-mining, preferring to develop its own far less well-endowed deposits. Romania clearly embraced a policy of expanding a heavy industrial base on largely nationalistic lines.

Kaser (1967, p. 49) indicates that following the Hollohaza conference, Czechoslovakia, the GDR and Hungary, who had already announced their five-year plans, published new plans. Romania's first plan was not approved until December 1950 and therefore accommodated the upward revisions, Bulgaria did not give details of the upward revision and Poland recast its more detailed plans in the light of changed requirements.

Details of the policy can be seen in Table 2.2. Industrial output

Table 2.2: First Five-Year Plans of the East European Countries (pre-plan output = 100)

	Bulgaria	*Hungary* (revised)	*Poland*[a]	*Romania*	*Czechoslovakia* (revised)
All industry	219	310	258	244	198
Group A	320	380	254	260–270[b]	233
Group B	175	245	211	225–210[b]	173
Steel[c]		275	230	357	179
Machinery	680	na	360	216	352
Metal working	266	490	300	220	231
Agriculture	159	154	150	188	137

Sources: Spulber, pp. 288, 290, 361. Also see notes. His source — original plans of country concerned.

Notes: a. Poland — six-year plan. b. Romania — no figures are provided for planned output of Group A and B. The estimates are based on the actual growth ratios of Group A and Group B from 1950 to 1953, scaled according to the growth rate for all industry (Academia, 1964, p. 88). c. Planned output at the end of the five-year plan divided by planned output at the end of the reconstruction plan.

was planned to grow faster in each country than agricultural output and within industry Group A was to grow faster than Group B, with substantial growth rates scheduled for the output of steel, machinery and equipment and metal working. In Czechoslovakia, arms production increased fourfold between 1950 and 1952 (Holesovsky, 1977, p. 709).

The priorities involved in the plans can be better illustrated by the size and structure of investment planned to achieve these growth rates. In Romania 51.4 per cent of all state investments in the first five-year plan were destined for industry with 82 per cent of that amount scheduled to producers' goods. Transport and communications were to receive 16.2 per cent, part of which was to restore the rail network while a major component was dedicated to the grandiose Danube-Black Sea canal which was eventually abandoned with only 10 per cent completed in 1953. Agriculture was planned to receive 10 per cent, while housing, education, health and scientific research actually received 11 per cent of total investment.

In practice, total investment in the socialist sector grew by an annual rate of more than 30 per cent between 1951 and 1953, industry received over 56 per cent of this investment, Group A more than 50 per cent of total investment and nearly 90 per cent of industrial investment. Investment in Group A in 1953 was three times its level in 1950 (Academia, 1958, pp. 23–4; Montias, 1967, pp. 24–5).

The initial draft of the Hungarian plan allowed for investment of 51 billion forints, of which industry was to receive 21 billion (42 per cent), of which 18 billion were to be directed to producers' goods. Agriculture was to receive 8 billion forints. The revised plan nearly doubled industrial investment to 40 billion forints, total investment was planned to increase by two-thirds to 85 billion forints, of which agriculture was to receive an additional 5 billion forints (Balassa, 1959, p. 32).

Official statistics indicate that the proportion of national income devoted to investment was around 20 per cent in each country in 1950 and 1955 (rising to 25 per cent from 1951 to 1953). Zauberman (1964, pp. 39–41), however, argues that this understates the real level of investment, partly because the value of consumer goods was inflated by increases in turnover taxes while the prices of capital goods were kept artifically low by subsidies. By estimating 1953 output and investment at 1949 prices the United Nations calculated the share of investment for Czechoslovakia and Poland at 31 per cent and 38 per cent respectively. Making additional corrections to measure Gross Capital Formation as a proportion of Gross National Product according to western conventions, Zauberman arrived at figures of 44 per cent and 48 per cent for Czechoslovakia and Poland in 1953 (compared with a figure of 18 to 20 per cent for Western Europe in 1950 and 1956).

In addition to placing a heavy strain on domestic consumption levels and food production, the autarkic nature of the policies pursued in each country reduced their economic complementarity, increased the supply of similar engineering and metallurgical products and increased the demand for energy and raw materials. In addition, Czechoslovakia diverted its production away from the higher quality engineering and consumer goods production for which it had established a pre-war reputation, towards armaments manufacture. As a result, the 'radial' pattern of trade developed in CMEA whereby the USSR met the energy and raw material demands of Eastern Europe and received heavy industrial products in exchange.

Following the death of Stalin, the 'New Course' in Eastern Europe placed greater emphasis on agricultural production and light industry. The more grandiose investment projects were terminated, the production of new steel capacity was delayed, and support for the Stalinist pattern of industrialisation in Eastern Europe was temporarily undermined.

Rakosi was attacked by Mikoyan, Khrushchev and Molotov for his agricultural policy and for the excessive build up of iron and steel production without a domestic raw material base (Fejto, 1974, pp. 38–9). In 1954 investment in Hungarian heavy industry was reduced by 41 per cent, investment in light industry increased by 12 per cent and in agriculture by 20 per cent (Balassa, 1959, p. 36).

But the political and economic structure of Eastern Europe had been essentially determined in the Stalin era. The 'New Course' was identified with Malenkov, who was overthrown by Khrushchev in 1955. Khrushchev still needed the support of conservative factions, in particular the 'metal-eaters' in the Kremlin, and reaffirmed the primacy of heavy industry. Nagy, who had replaced Rakosi as Prime Minister in Hungary and implemented the more consumerist policies, was himself ousted and Rakosi resumed the position of Prime Minister as well as Party Secretary.

Khrushchev's destalinisation speech in February 1956 and the ensuing disturbances in Poland and Hungary, the temporary reinstatement of Nagy, and ultimately the Soviet invasion of Hungary clarified the limits to the degree of destalinisation and reform that would be tolerated by the Soviet leadership and the orthodox communist leaders in Eastern Europe. In Poland and Hungary, Gomulka and Kadar emerged as leaders who appeared capable of steering a course between the demands of the population and the need to preserve Party authority.

Equally critically the pursuit of autarkic policies and the principle of the priority of heavy industry in each country of the bloc had been established. The economic effects of this can be illustrated by Romanian experience (Table 2.3). In 1954 capital investment in the Romanian economy was cut by 1,588 million lei (12 per cent). Investment in industry, Group A, was cut by 884 million lei, but investment in Group B increased by 308 million. Investment in machinery and equipment was cut by 629 million but 595 million of this cut came from imports, while investment from domestically produced machinery and equipment stabilised. When higher levels of investment and the shift back to group A were resumed in 1955, the entire increase in demand for machinery and equipment was met by domestic production. Imports from

Table 2.3: Capital Investment in Romania, 1950–6
(million internal lei; constant 1955 prices)

	1950	1952	1953	1954	1955	1956
Total	5,650	10,587	13,463	11,875	13,178	14,784
Industry	2,728	5,905	7,678	7,102	7,582	8,264
Group A	2,365	5,246	6,960	6,076	6,331	7,276
Group B	362	659	717	1,025	1,252	988
Value of machinery and equipment installed	2,032	3,549	4,413	3,784	4,279	4,463
(i) Produced domestically	1,031	2,109	2,800	2,766	3,180	3,419
(ii) Imported	1,001	1,440	1,613	1,018	1,099	1,044

Source: Academia (1958), pp. 23–7.

Czechoslovakia, which trebled between 1950 and 1953 fell back to their 1950 level in 1954 and declined further in 1955 and 1956. This pattern was to have profound consequences both for Czechoslovak economic development and for economic development in the bloc.

3 THE PLANNING SYSTEM IN EASTERN EUROPE

The Planning System and Economic Growth

The outstanding achievement of the Stalinist system of central planning is that it allowed the economy to operate at a high level of recorded activity over prolonged periods, resulting in a rapid rate of growth of net material product and industrial output in particular. This was achieved by devoting over a quarter of national income to investment which simultaneously increased the productive capacity of the economy and maintained a high level of aggregate demand. The industrial labour force was augmented by the (forced) transfer of labour from agriculture, where marginal productivity was considered to be low, and by the increased participation of female workers in the state and co-operative sectors of the economy.

Central controls over the supply of consumer goods prevented excess consumer demand diverting resources away from investment and, as the system was modified on an ad hoc basis in the 1930s, improved controls over the banking system and central controls over prices and wages helped to combat open inflation. Similarly, the state monopoly of foreign trade prevented consumer demand from spilling over to imported commodities and enabled the state to concentrate imports on machinery and equipment and raw materials not available domestically (notably non-ferrous metals).

This model of development which depended on generating economic growth mainly by increasing the quantity of inputs available to those sectors of the economy producing material output was subsequently termed 'extensive growth' by Czechoslovak reform economists who contrasted this approach to 'intensive growth' which placed greater emphasis on improved efficiency of the utilisation of existing resources. (See Chapter 4.)

It might be expected that 'pure extensive growth' would increase the productive capacity of the economy and output per head of the population by utilising a larger proportion of the labour force that is available for employment, but that this would require the employment of increased quantities of labour in less efficient areas where output per worker is below the average for the economy as a whole, which would in turn decrease the average level of output per worker employed. In practice, however, recorded output per employed worker in industry has actually increased both in the USSR in the 1930s and throughout the CMEA area in the post-war period, until the late 1970s.

In the USSR in the 1930s this was achieved, not by modernising existing factories, but by constructing new large-scale plants incorporating modern technology much of which, as Sutton (1971) has demonstrated, had been developed outside the USSR itself and had been imported or copied and/or installed with the assistance of foreign technicians. Consequently, the output of the new plant helped to raise the average productivity per employed worker. As much of the new plant was capital-intensive while the USSR possessed considerable surpluses of unskilled labour, labour-intensive processes coexisted (frequently on the same site) with capital-intensive processes. Thus, as the rate of investment in industry and the growth of the industrial capital stock grew faster than the industrial labour supply, labour could, after training, be transferred to more productive uses resulting in further increases in output per employed worker, purely by replicating and diffusing existing technology, although a stage would be reached when optimal factor proportions had been attained and further increases in output per worker would require improvements in technology and/or economic organisation.

Although the process would not be (and has not been) as mechanical as that outlined above, the implications of this stage of development are critical. Once reached, the rate of growth of output per worker will only increase at a rate determined by the ability of the economy to generate its own technology, or acquire technology from overseas and/or utilise available resources more efficiently. This, in turn, has considerable implications for macro-economic control. Unless non-consumption expenditure is reduced, or a balance of payments deficit is incurred, real wages and incomes can only grow as fast as the overall growth rate, which will require a reduction in the rate of growth of money wage rates if inflationary pressures are to be avoided. Technical innovations and improvements in economic organisation may necessitate closing down less efficient plants and activities in order to release resources (including labour) for use elsewhere. As the process of labour transfer is unlikely to be instantaneous this will in turn increase the tendency towards 'frictional' unemployment.

The rate of growth of Soviet industrial output slowed from over 10 per cent per annum to under 2 per cent between 1937 and the beginning of the war (Kaplan and Moorsteen, 1960), although this must in part be attributed to the purges which had eliminated many of those with managerial skills and the general debilitating effects of sanctions and coercion as methods of motivation, while the labour camps themselves were beginning to impose costs on the economy by reducing the labour available for more productive uses.

In the immediate post-war period (1945–50) the rate of growth of industrial output exceeded 20 per cent per annum as the USSR reconstructed its war-damaged economy and was able to incorporate technical developments that had taken place in Europe and the USA from which it had been effectively isolated during the war period (Gomulka, 1976). A more normal growth pattern was resumed once this 'stock' of technical developments and the scope for reconstruction had been assimilated, and the effects of the cold war on international technology transfer were felt.

The pattern of a rapid burst of growth of industrial output, exceeding 10 per cent per annum (according to official statistics), resulting from high rates of growth of investment and of the industrial labour force was repeated in the first five-year plan period in each of the East European CMEA nations, although the growth rate of industrial output declined in each country, except Bulgaria, in the second five-year plan period. A downward trend in the growth rate of industrial output (and hence net material product) can be detected in each country of the bloc throughout the period, resulting in a negative rate of growth of industrial output in Poland and Hungary in 1980 (Table 3.1).

Although there has also been an absolute decline in the industrial labour supply in Hungary, Poland and Bulgaria and a cut in the annual flow of investment (but not necessarily the stock of industrial capital) in Czechoslovakia, Hungary and Poland (Table 3.2) at the end of the 1970s, the initial decelerations of the growth of industrial output were not accompanied by *equivalent* reductions in the rate of growth of inputs to industry. This factor was reflected in increasing incremental capital-output ratios throughout the bloc (ie, an increasing growth of capital was required to bring about an equivalent increase in output).

This fact, combined with the evidence from Tables 3.1 and 3.2, lend some *prima facie* support to the hypothesis, advanced by the advocates of economic reform, that the model of extensive growth

Table 3.1: Annual Average Growth of Net Material Product and Gross Industrial Output

	1951 –55	1956 –60	1961 –65	1966 –70	1971 –75	1976	1977	1978	1979	1980
Net material product										
Bulgaria	12.3	9.7	6.7	8.7	7.8	6.5	6.3	5.5	6.6	5.6
Czechoslovakia	8.2	6.9	2.0	6.9	5.7	3.8	4.4	4.2	2.7	3.3
GDR	13.1	7.2	3.4	5.2	5.4	3.8	5.2	3.5	3.4	4.6
Hungary	5.8	6.0	4.1	6.8	6.2	2.9	8.5	4.6	1.9	–0.7
Poland	8.6	6.6	6.2	6.0	9.4	6.8	5.3	2.8	–2.2	–5.6
Romania	14.2	6.6	9.0	7.8	11.3	9.9	8.4	7.8	5.8	3.0
Gross industrial output										
Bulgaria	13.7	15.9	11.7	10.9	9.1	6.8	6.8	6.9	5.4	4.2
Czechoslovakia	10.9	10.5	5.2	6.7	6.7	6.5	5.6	5.0	3.7	3.5
GDR	13.7	8.7	5.8	6.5	6.5	5.9	4.8	4.7	4.5	4.7
Hungary	13.2	7.6	7.5	6.2	6.4	4.6	6.6	4.9	3.0	–1.6
Poland	16.2	9.9	8.4	8.3	10.4	9.3	7.5	4.8	2.1	0.0
Romania	15.1	10.9	13.8	11.9	12.9	11.5	12.7	9.0	8.2	6.5

Source: Calculated from *Statisticheskii Yezhegodnik Stran-Chlenov SEV*, various years. The Government Report on the State of the Economy for Poland, July 1981 gives a lower series for industrial output for 1976 to 1980.

Table 3.2: Annual Rate of Growth of Inputs to State and Co-operative Industry

Labour (Annual Average Number of Employees)

	1951 –55	1956 –60	1961 –65	1966 –70	1971 –75	1976	1977	1978	1979	1980
Bulgaria	5.2	10.9	4.4	3.9	2.3	1.0	0.7	1.0	1.3	–1.3
Czechoslovakia	10.4	3.4	1.9	1.2	0.8	0.1	0.4	0.6	0.5	0.3
GDR	6.6	1.6	0.6	0.6		1.1	1.1	0.7	0.7	0.2
Hungary	10.5	3.4	2.8	2.6	0.3	–1.2	–0.9	–0.3	–1.6	–2.6
Poland	6.6	2.1	3.3	3.3	3.1	0.3	1.0	–0.2	–0.3	–0.3
Romania	5.9	3.0	5.9	4.3	6.3	3.8	4.1	2.6	3.9	3.2

Investment (constant prices)

	1951 –55	1956 –60	1961 –65	1966 –70	1971 –75	1976	1977	1978	1979	1980
Bulgaria	14.0	15.3	14.0	12.7	6.0	1.4	17.6	1.3	0.0	8.6
Czechoslovakia	4.0	16.6	3.4	4.6	7.1	6.2	1.6	3.6	6.6	–0.5
GDR	17.3	16.8	7.1	8.9	4.0	8.3	6.8	5.1	3.9	3.7
Hungary	5.1	11.4	5.9	10.5	9.1	8.1	12.3	3.3	–2.2	–12.1
Poland	13.8	7.3	8.0	7.6	22.0	3.0	–0.7	–2.1	–14.2	–12.7
Romania	22.9	8.8	13.8	11.2	12.4	2.2	16.3	20.2	7.7	2.5

Source: As for Table 3.1.

is best suited to countries at a low or intermediate level of economic development in which there is a substantial agricultural population capable of learning industrial skills and of transferring to industrial employment, and is less capable of sustaining the growth of a more mature economy.

The highest rates of growth of industrial output and net material product have been sustained in the least-developed countries, Romania and Bulgaria, who have also maintained the highest growth rates of industrial investment. Both Czechoslovakia and the GDR, which ran into difficulties in increasing the industrial labour force in the late 1950s and early 1960s, experienced not only the lowest rates of growth of industrial output and net material product in the 1961–5 plan period, but also the sharpest decline in the growth of industrial output per worker and succeeded in reversing these trends by increasing industrial investment rates in the 1966–70 plan period. Poland only succeeded in reversing the downward trend in growth rates in the 1971–5 plan period by massive imports of western machinery and equipment which permitted an increase in industrial investment of 22 per cent per annum. A slight improvement in Hungarian economic performance can be detected in the period following the launching of the New Economic Mechanism in 1968. All the countries have suffered from a substantial deceleration of growth rates in the 1976–80 period following the increase in world energy prices which has required cutbacks in domestic investment rates.

The best test of the long-term viability of the extensive growth model may be provided by the experience of Romania, which has pursued the policy most consistently throughout the period, was not so immediately affected by the initial impact of the 1973 energy crisis (being a net oil exporter at the time) and has not diverted its domestic resources towards developing bloc energy projects located in the USSR.

Official statistics show that the proportion of national income devoted to investment has risen in each five-year plan period from 16 per cent in 1956–60 to 36.3 per cent in 1976–80. Over 45 per cent of total investment has been concentrated in industrial producer goods, while agriculture and industrial consumer goods have received approximately 13 per cent and 7 per cent of total investment respectively. As a result, therefore, 13.3 per cent of net material product was devoted to investment in industrial producer goods in the 1976–80 period. Simultaneously, the growth rate of both industrial output and net material product have declined in each five-year plan period (except 1956–60) and in each year since 1975. Similarly the growth rate of output per employee in industry has declined in each five-year plan period and despite attempts to boost the industrial investment rate in 1976 and 1977 has continued to decline.

Thus, despite impressive growth rates of industrial output and net material product Romanian statistics provide clear evidence of a declining capital-output ratio through time with a continuous decline in the growth of industrial output per man being associated with an increase in the growth rate of capital inputs to industry. It is difficult, however, to know whether this should be ascribed to the economic system or whether a similar economic performance might have been achieved with a different economic system. Other economies at intermediate levels of development and with substantially different economic systems (Greece, Spain, Japan, Taiwan and more recently South Korea) have attained growth performances comparable to those attained by Bulgaria and Romania, largely through the aegis of international technology transfer (Gomulka, 1971), while Yugoslavia has achieved a comparable growth performance with a decentralised socialist economic system. Similarly most western industrialised nations have experienced a deterioration in their rates of economic growth through time and have been adversely affected by the oil crisis. Finally, although the Asian socialist countries (Mongolia, North Korea, Vietnam) have adopted a centralised system of development, Soviet specialists are encouraging African countries in the 'new communist Third World' (Angola, Mozambique) to pursue a less centralised policy more akin to NEP (Wiles and Smith, 1982).

On balance, it appears that the Soviet system of central planning has not prevented, and has probably facilitated, a policy of rapid industrial growth for countries at an intermediate level of development by taking up underutilised productive potential. This has been largely achieved by providing machinery and training to labour with low marginal productivity. As the potential for further substantial increases in output of labour has declined, the growth rates of output and output per worker have also declined. The implications for future growth in Eastern Europe are similar to those reached for extensive growth in the USSR — further growth will depend on the ability of the economy to generate new technology, and/or acquire improved technology from socialist or non-socialist countries and diffuse it throughout the economy, and/or stimulate improvements in economic organisation.

The remainder of Part I will examine the operation of central planning in Eastern Europe at the enterprise level and attempts to reform enterprise operation, particularly with a view to seeing whether the centrally-planned economies possess a systemic bias against innovation at the enterprise level. The impact of this analysis on households and macroeconomic control is examined in Part II.

Finally, the operation of extensive growth and the Soviet economic system in Eastern Europe has been complicated by a number of factors relating to foreign trade and economic relations with the USSR itself which are examined in Part III. The most important of these factors should however be considered when examining other aspects of economic operation and may be indicated at this stage:

(i) The model of extensive growth puts greater strains on the demand for raw materials, energy inputs, and so on. The USSR is well endowed with most major raw materials and could satisfy its basic demands from internal sources. The East European economies are in the main resource poor and must acquire raw materials either from the USSR or from non-socialist sources.

(ii) The USSR and the less-developed CMEA economies have been capable of absorbing and adapting existing technology generated outside their economies (frequently from within the CMEA itself). The more technically advanced East European countries are more dependent on extra-CMEA contacts to improve domestic technology and are more vulnerable to western restrictions.

(iii) The predilection for large-scale plant combined with small domestic markets has increased the dependence of East European economies on exports to maintain satisfactory production runs.

(iv) The greater dependence on foreign trade has made the problem of rational calculation more difficult, particularly where the cost of domestic production must be compared with the cost of production in other socialist economies and in world markets.

(v) The cost of wrong large-scale investment decisions will be proportionally greater for smaller economies and may be *actually* greater where intended exports to market economies cannot be absorbed domestically. Although it may be economic to continue production from such plants for export purposes, if variable costs are covered, market economies may be reluctant to accept the 'sunk costs are sunk costs' argument and impose restrictions on such exports.

The Operation of the Planning System and the Industrial Enterprise in Eastern Europe

A considerable amount of research has been conducted in the West to examine the performance of the planning system and the industrial enterprise in the USSR and Eastern Europe. As a result a model of the microeconomic operation of those economies has been developed which commands a broad degree of consensus amongst western analysts and which has been confirmed as reasonably accurate by East European sources and by Soviet and East European emigré economists.

The basic principles of the operation of the Soviet enterprise in the Stalinist period were first examined by Berliner (1957) on the basis of extensive interviews with Soviet emigrés with managerial experience, while Nove (1980) has amply demonstrated the problems of the complexity of planning an entire economy in the detail attempted by Soviet planners. The methods of planning and the problems of communication and aggregation have been analysed by Ellman (1971, 1973), and Birman (1978), a Soviet emigré economist, has provided detailed confirmation of western models of the operation of the Soviet economy.

Studies by East European economists with personal experience of the operation of the planning system, such as Kornai (1959), Sik (1967), Selucky (1972) and Brus (1972), have confirmed that the model was transferred in a broadly unaltered form to Eastern Europe. Granick (1975) has undertaken the most detailed appraisal of enterprise operation and managerial behaviour in Eastern Europe on the basis of interviews conducted with ministerial officials, enterprise managers and technical personnel in Hungary, the GDR and Romania in the early 1970s (ie, in the immediate post-reform era). His studies confirm that the Soviet economic model remained virtually unaltered in Eastern Europe (except in Hungary after 1968), but question the validity of the 'success indicator' theory of suboptimisation outlined below as an explanation of managerial behaviour in the GDR and Romania. Recently a number of studies of the operation of the planning system in individual East European countries have been conducted (Jeffries, 1981; NATO, 1980).

There are therefore reasonable grounds for supposing that a general model describing the behaviour of the planning system and enterprise operation, although a considerable abstraction, can be usefully applied to all the CMEA countries. Furthermore, as I shall argue that (following Brus, 1979), with the possible exception of Hungary, the economic reforms enacted in the mid 1960s largely involved changes in the 'rules of the game' and left the basic economic system unaltered, the model developed here (derived from all the above sources) can still provide a valid basis for understanding the operation of the East European CMEA economies.

Central planning consists of two interrelated aspects — plan formulation and plan implementation. The success of the former depends on the quality and accuracy of the information received by planning bodies and the success of the latter requires the development of policy instruments which will harmonise the interests of production agencies with those of the planners.

An omniscient, omnipotent central planning agency with full knowledge of all the economic variables at its disposal and with the ability to draw up optimal plans, to issue precise instructions for their implementation and to supervise their fulfilment is a practical impossibility, even with sophisticated techniques of calculation.

Consequently, planners lower their sights to construct feasible, rather than optimal, plans and the process of planning has to be divided up between participants in the economic process which can lead to problems resulting from failures of communication. During the 1930s the Stalinist system evolved into a hierarchical planning structure in which information flowed upwards from enterprises through intermediate agencies to the Central Planning Agency and plan instructions flowed back in the opposite direction.

Although planning and production are conducted by state agencies, they are responsible at all stages of the process for implementing Party policy. The State Planning Committee (SPC) is responsible both for the broad execution of macroeconomic policy (eg, determining the proportions of national income to be devoted to consumption and investment and the sub-division between private and social consumption, investment in producers' goods, consumer goods, defence), and for the more detailed implementation of microeconomic equilibrium.

The broad directions of Party policy are established in perspective plans, covering a 10–15 year period, and five-year plans which specify growth rates of production for individual sectors of the economy, and which establish the basis for investment programmes. Such plans do not contain instructions to enterprises and as such are not 'operational'. We may distinguish at this stage between the functions of central *planning* (ie, drawing up the guidelines for future periods in order to achieve centrally-determined objectives) and central *direction* or administration of the economy (ie, issuing detailed operational instructions to enterprises telling them what to produce, from where they will receive their inputs, and so on). The critical distinguishing feature of the Stalinist system is that it is a centrally *directed* economy (Birman, 1978; Holesovsky, 1980).

The operation of the system may be described by the construction of annual plans (although even this is an overoptimistic period over which planners can control events in detail, particularly where foreign trade is concerned).

The SPC draws up material balances which specify the available supply and distribution of (demand for) a highly aggregated group of commodities (numbering 1,500–2,000 in the USSR). This task is facilitated by information concerning past production levels and demand for resources, and knowledge about new plant and capacity coming on stream and changes in the available supply of raw material, labour, and so on. Birman (1978, p. 161) argues that without such information on past production levels the process of planning would be 'unbelievably difficult' and that 'colossal blunders' were made when Soviet specialists helped draw up the first five-year plans for the East European countries. In practice, Birman argues, the process of plan formulation is based entirely on the 'from the achieved level principle' — the planners draw up plans by adding a certain percentage growth to achieved output targets in order to stimulate production agencies to produce more.

Aggregated output targets are then handed down to Industrial Ministries which are responsible for their execution within a broad sector of industry (eg, heavy engineering). From there, they are passed on to sub-departments responsible for their execution in a more narrowly-defined industrial sector, who in turn pass the targets (sometimes through intermediary agencies) to the enterprises under their supervision. At each stage of the process a greater degree of detail is added to the instructions passed down to lower authorities (disaggregation).

As the same agencies were responsible for the upward flow of information on which plans were based, the downward flow of instructions is also based on the achieved level principle — each agency instructs its subordinates to achieve a percentage more than they claimed to have achieved in the previous plan period (Birman, 1978, p. 162).

Finally, each enterprise receives a detailed annual operational plan, containing physical instructions which may specify in considerable detail the outputs the plant is to produce (and to whom they should be delivered), the material inputs it is to receive (and from whom they should be received), the quantity and type of labour to be employed and wage rates to be paid, new technical innovations to be used, and so on.

A hierarchical structure also operates within the enterprise, with the enterprise director responsible for the implementation of plans in general and line managers responsible for the fulfilment of tasks within their jurisdiction — they are, however, subject to a number of internal and external checking agencies including ministerial superiors, local Party organisations, trade unions and the State Bank. A major (compromise) feature of the Stalinist economic system is the use of monetary relations and individual enterprise accounting (*Khozraschet*) as a means of exercising detailed operational supervision over the activities of enterprises to ensure that they operate according to plan instructions. Each physical instruction is complemented by a financial instruction contained in the enterprise financial plan, which will specify such items as: a revenue plan composed of the value of planned outputs multiplied by their wholesale prices; a material input cost plan composed of the volume of planned inputs multiplied by wholesale prices; an investment cost plan detailing payments connected with centrally-determined investments; a labour cost plan which could specify in considerable detail the amount of labour to be employed and wage rates to be paid; a taxation plan; a profits plan composed of the difference between planned revenue and planned costs, and detailing its distribution between payments to the state and retention for incentive funds, and so on.

The State Bank is in a unique position to exercise detailed micro-economic control over the enterprise and to supervise all aspects of its operation by ensuring that financial payments are only authorised for transactions established in the plan.

The system of enterprise stimulation also relies heavily on monetary incentives. Enterprises have three basic types of incentive funds: a material incentive fund for the payment of monetary bonuses to the work force (managers' bonuses are occasionally calculated on a different basis); a socio-cultural fund for financing enterprise social consumption (for example, canteens, sports and holiday facilities); and a production-development fund for financing small-scale capital expenditure (Ellman, 1971, pp. 131–62; Jeffries, 1981, p. 6). Bonuses are paid out for the successful fulfilment or overfulfilment of plan targets; in addition, however, the bonuses of ministerial personnel are also attached to the successful implementation of plan tasks by enterprises under their control. The orthodox model of enterprise behaviour, largely due to the research of Berliner (1957), proposes that enterprise managers are strongly motivated by the pursuit of bonuses, which form a substantial component of their current income, and that the consistent procurement of bonuses is viewed by superior authorities as an indication of managerial competence which will favourably affect longer-term career prospects.

It is at this stage that problems of communication, information and aggregation become significant. Enterprise managers can affect the plans they receive and hence their ability to secure bonuses by the information they provide upwards to planning authorities. They are therefore strongly motivated to aim for slack (easily attainable) plans and both the workforce and ministerial superiors are motivated to accept this process.

Consequently, the information received by central planners is frequently a poor guide to enterprises' true production potential. Central planners attempt to counteract this by imposing taut plans on the enterprise which will stimulate a higher level of output. Informational uncertainty and the attempt to impose taut plans prevents central planners from issuing precise, accurate instructions to enterprises as a result of which many of the plan indicators are contradictory. The fulfilment, and in particular the overfulfilment, of one target (eg, output) may require the violation of a constraint (eg, material inputs).

Under these circumstances, enterprises will attach most importance to the fulfilment and overfulfilment of targets to which bonuses are linked (which presumably central authorities consider to be the most important, and which Nove (1980, pp. 96–102) calls 'success indicators') at the expense of other targets. The process of plan formulation requires that a large number of heterogeneous commodities are aggregated according to some common denominator. For many items (eg, furniture composed of beds, tables, chairs), the most reasonable denominator for aggregation may be monetary units. For other items (eg, coal) a physical denominator such as weight may be used, or alternatively an indicator specifying the number of units (eg, cars) to be produced.

Bonus maximisation provides enterprises with an incentive to act according to the letter, rather than to the intention of the plan and to undertake activities that are socially irrational. In the unreformed Stalinist system, priority was given to the fulfilment of gross output targets, and consequently the method of aggregation critically affected the plant's output. If, for example, outputs were specified in terms of weight, enterprises had an incentive to concentrate their output on heavy items within their range (pre-cast concrete slabs were produced predominantly in larger units from which a higher weight output could be achieved with a given labour force) or to produce low-quality outputs (coal output could be exaggerated by the inclusion of non-combustible material). If outputs were specified in volume terms (numbers) enterprises had an incentive to produce a larger number of smaller items (eg, small nails).

Two points may need consideration. Firstly, East European officials are well aware of these drawbacks and it is from their criticisms that such examples (which cause much mirth to western commentators) are derived. Such violations of the spirit of the plan are similar to the concept of the 'professional foul' — they occur when the rewards for violations outweigh the sanctions. Planners may, and frequently do, change the reward system to render such activities unprofitable. On the other hand, changing the method of calculating the success indicator (eg, from weight to volume) or the success indicator itself (from gross output to profit) *may* result in the enterprise taking more socially rational decisions, or may purely result in another form of violation of the planners' intentions.

The checking agencies may do little to prevent socially irrational activities. Ministerial authorities whose bonuses are linked to the same criteria as enterprises retain a vested interest in enterprise bonus maximisation, while local Party officials may be reluctant to jeopardise bonuses for the workforce. Furthermore, ministerial and Party officials may find themselves blamed if an enterprise under their jurisdiction fails to meet its major indicators and may actively participate in the processes they are supposed to prevent (Andrle, 1976; Berliner, 1957). Even the State Bank appears to be reluctant to take action that would effectively bankrupt an enterprise or cause it to fail to meet output targets and have to make plant and workers redundant (Wiles, 1968, p. 48).

Granick (1975) questions the validity of the 'success indicator' model as an explanation of socially irrational managerial behaviour in the GDR and Romania. In Romania, bonuses constituted a small proportion of managerial income at the time of Granick's interviews, enterprise plans were not considered to be excessively taut, and planners felt that the inability of an enterprise to meet plan targets reflected badly on the plan-setting abilities of the planners themselves, and made frequent adjustments to enterprise plans during the course of the year to ensure that plans were actually achieved, thereby nullifying the effects of managerial incentives linked to the fulfilment of plan drafts. This was largely a reflection of the fact that many economic decisions were taken at a level higher than the enterprise itself (Granick, 1975, pp. 89–127). In the GDR, bonuses constituted a larger proportion of managerial income, but were not so rigidly tied to the successful implementation of a single aggregated index, and managers had far greater freedom of operation including the selection of the product mix, the choice of customers, raw materials and even in the determination of the composition of the labour force and wage rates (Granick, 1975, pp. 210–11).

Granick's findings place less emphasis on bonus maximisation *per se* as a cause of socially irrational behaviour and require us to look to the nature of the economic system and, in particular, to the relationship between the enterprise and superior authorities for an explanation. Indeed, the term 'enterprise' is a misnomer for what is effectively a centrally-directed production agency which tends to be regarded, in Nuti's words as a 'sausage machine transforming inputs into outputs according to given production functions . . . rather than as a distinct organism living in a specific environment' (Nuti, 1981b, p. 61). Enterprises do not own the assets they utilise and cannot sell them to other enterprises that may be able to put them to better use, or raise income to acquire other assets. Their areas of production are narrowly defined, product lines are determined by higher authorities and they cannot diversify into more profitable (and/or more socially advantageous) fields of production on their own initiative. Enterprises can only be founded, amalgamated and disbanded by central authorities.

If enterprise managers conduct their activities and their production plans in order to satisfy superior authorities (including the Party) rather than to meet the needs of end users of the product, socially irrational activities may result from informational uncertainty. Central authorities cannot always know what end users require and cannot specify their intentions accurately and precisely. Enterprise managers must therefore interpret central planners' intentions, but when in doubt may tend to act according to the letter of central instructions rather than the spirit.

Kornai's experience (1959, pp. 118–22) in Hungarian industry, however, indicated that managers do attach greatest importance to the fulfilment of indicators on which their own bonuses are based but that material incentives, moral incentives and sanctions tend to support the system of instructions which may not only be contained in plan documents but passed on through less formal channels. On occasions, the pursuit of bonuses may even conflict with moral pressures, or direct instructions, and the manager will be required to react accordingly. Enterprise behaviour will therefore be largely determined by the environment in which the manager operates, which includes the *existence* of a system of instructions, as much as a system of bonuses.

Provided central authorities attach greater importance to maintaining production and maximising the growth of gross material output, and the Party attaches considerable importance to maintaining full employment and in particular to avoiding frictional unemployment, these major goals will be communicated to enterprise managers, whatever the system of bonuses or contradictory instructions. These goals were reinforced in the unreformed Stalinist system by linking managerial bonuses to gross output, while enterprise bonuses were linked to gross output and multiplied by the labour force.

The cult of gross output has resulted in many well-documented abuses (Sik, 1967, Chapter 1; Kornai, 1959, Chapter 3). Enterprises are not motivated to estimate whether the value of end products outweighs the costs of production or to seek low-cost solutions to centrally-determined problems. Inputs appear to be costless to the enterprise, and so in the process of plan formulation they overindent for labour, capital and raw materials; in the process of plan implementation they tend to accept any inputs allocated to them by the supply system, even if they are surplus to immediate requirements, or not ideally suited to the job in hand, in the hope that they may help to realise output or may be unofficially bartered for a more useful product. Enterprises hoard inputs (especially labour) in anticipation of future demands that may be put upon the enterprise. A sellers' market arises in which the official supply system is marked by shortages while stocks of surplus commodities may exist within enterprises themselves. An (officially tolerated) network of 'fixers' moving between enterprises and bartering and/or buying and selling surplus products arises and is effectively essential to the operation of the system. Enterprises ensure against supply uncertainty by building small low-capacity workshops for vital inputs, and so on.

Enterprises striving to meet gross output targets will tend to ignore the quality of output, knowing that under sellers' market conditions whatever is produced may be accepted by the state-warehouse system or by the state-retail network and will contribute to plan fulfilment. Problems of aggregation will result in enterprises concentrating on production runs of outputs that most easily fulfill specified indicators (and engage in the practices linked to high weight or high volume output indicated above) and they will produce with a very low range of variation of sizes, styles, and so on (assortment problem) and may even continue to produce commodities after the market for that output has been satisfied.

Finally, spurts of economic activity towards the end of the planned period (storming) resulting in excessive overtime, shoddy production and high-cost solutions to achieve plan targets, followed by a lull in activity at the beginning of the next period, are reflected in data which indicate peaks in production in the last ten days of the month, in the last month of the quarter and in the last quarter of the year. Kornai (1959, p. 139) suggests that storming largely applies to the fulfilment of indicators linked to material incentives and their periodicity indicates that the bonus system is responsible.

A further critical problem, that the nature of organisation of the Soviet enterprise hinders technical innovation, will be examined in the next chapter.

4 INNOVATION AND ECONOMIC REFORM

A society that centralizes investment decisions may be expected to be biased against innovations whose introduction cause losses to existing operators by, eg improving the design and quality of current output and by increasing its variety through the introduction of substitute goods . . . there would be no similar reluctance to undertake the production of entirely new articles not meant as substitutes for any one existing good, or of capital goods whose eventual impact on existing producers is hard to evaluate.
Albert O. Hirschman, *The Strategy of Economic Development*

Innovation and the Centrally-Planned Economy

A major feature of the centralised economic system has been that it has overcome 'man's myopic faculty' and maintained a higher rate of investment over a longer period of time than has typically been established by market economies dependent on the individual actions of households desisting from current consumption for the supply of savings, and on corporations deciding by how much to increase their productive capacity for the demand for investment. Such investment rates have been largely unaffected by (self-realising) anticipation of future purchasing power and fears of future recession.

It might be expected that the spreading of risk over the economy as a whole would permit the implementation of bolder, more imaginative projects that would be beyond both the financial and planning capabilities of an individual entrepreneur, and that such investment would be concentrated in projects with a high social rate of return. It might also be expected that the centralisation of investment decisions, including the centralisation and concentration of research facilities, would permit new technological breakthroughs to be achieved and be rapidly diffused throughout the economy.

Soviet successes in some sectors of the economy that have received high priority — including certain aspects of the defence sector, an initial but forsaken lead in rocket technology, the scaling-up of standard technologies and their application on a large scale, certain aspects of the transportation of energy, such as electricity transmission (Hanson, 1981a, Chapter 3) — appear to support such expectations. It is noticeable that many Soviet successes have been achieved in the 'collective goods' sector, where the cost per head of the population is lower, and/or where the possibilities for substantial economies of scale exist. It is therefore questionable whether similar benefits could be achieved in the smaller East European economies, unless East Europe is viewed as a single entity governed by a common plan, or research and development resources are pooled on a bloc scale in order to achieve common objectives which would require at least voluntary acquiescence to supranational aims. The political and economic power of the USSR, combined with the location of energy and raw materials, would inevitably mean that either solution would increase East European dependence on the USSR.

Furthermore, many of the achievements of the Soviet economy have been matched by the USA without centralisation of decision-making on a comparable scale. This raises the question of whether the system of detailed operational planning offers the economy any dynamic benefits that could not be obtained with a less centralised system.

Hanson (1981a, pp. 51–2) argues that the successful implementation of central priorities is largely a question of national strategy and can be achieved (presumably if the will exists) just as capably by a government operating in a market economy on the basis of government contracts (eg, the US Space Programme and motorway construction in the USA and post-war Western Europe). Similarly, Brus (1975, p. 9) proposes that the benefits of centralised investment described above can be achieved by the 'state as entrepreneur' in a market economy.

The crucial dynamic advantage offered to Eastern Europe by the adoption of the centralised system of investment is the ability to influence macroeconomic aggregates and sustain high investment ratios over prolonged time periods, while the system of detailed operational instructions to enterprises may actually hinder innovation at the enterprise level.

Schumpeterian theories of innovation (Schumpeter, 1974 edition, Chapters 8, 9) suggest that the process of 'creative destruction' generated by competitive pressures provides the fundamental spur to technical development in a capitalist economy. The capitalist entrepreneur is stimulated to innovate in the production of new commodities and new methods of production (the creative element) by the lure of high profits which may even be sustained by the establishment of a monopoly. Such innovations will make pre-existing methods of production, or the production of inferior commodities, unprofitable and will cause their production to cease, possibly resulting in firms becoming bankrupt and workers being made redundant (the destructive element).

Even if products or processes can be monopolised, the lure of higher profits will cause new entrants to the market to produce similar, or even technically superior or lower cost products that do not conflict with existing patents, or even to produce completely new products that may destroy apparently secure markets. The fear of being 'innovated against' prevents existing producers from sitting back on their achievements and requires them constantly to improve their processes and their products, 'competition . . . acts not only when in being but merely when it is an ever-present threat. It disciplines before it attacks' (Schumpeter, 1974 edition, p. 85).

The process of creative destruction, in Hirschman's analysis, effectively internalises the benefits of innovation (through higher profits) and externalises the costs.

The implications of this analysis from the viewpoint of static welfare economics is that competitive pressures in a capitalist economy provide an excessive stimulus to innovate, which may be observed in a plethora of new, but similar (and on occasions technically incompatible) consumer products.

This situation is effectively reversed at the enterprise level in a centrally-planned economy. An enterprise that spontaneously attempts to produce a new product, or develop a new production process, is effectively attempting to do something it has not been instructed to do. It must therefore either seek to have its plan instructions altered (including output targets, sources and types of inputs, and possibly even its labour force plan) or must violate its plan instructions. As both moral and material incentives are linked to the fulfilment of existing plan indicators, violation of existing plans will result in the failure to secure bonuses and the possibility of reprimand.

Even to attempt to secure changes in plan instructions will involve enterprise management in haggles with ministerial officials, material supply agencies and other enterprises (who may resist attempts to make the necessary alterations to their own input and output targets, which could in turn affect their ability to fulfil plan targets). Local Party officials may be expected to resist attempts to change production lines that will cause a reduction in demand for labour.

As ministerial material and moral incentives are frequently linked to the fulfilment of aggregated targets by enterprises under their jurisdiction, permission for a change in an enterprise's product mix may still be conditional on the fulfilment of aggregated output targets. Problems of changing production lines, retraining workers, overcoming unanticipated problems, even the failure to receive a changed input mix from suppliers, may all result in cutbacks in current production and jeopardise the fulfilment of aggregated targets. If all these problems are overcome and an innovation is successfully brought into series production, the ratchet principle will ensure that future plan targets are pushed up to the highest achieved level.

There is therefore a strong disincentive for an enterprise to attempt to innovate; effectively the costs of innovation are borne by the innovator (internalised) and the benefits accrue to society (externalised) and there is correspondingly an incentive to sit back and fulfil existing instructions — Nyers has referred to this practice as the 'drowsiness' of socialist enterprises (Wiles and Smith, 1978, p. 91). Sik (1967, p. 88) attests to the fact that Czechoslovak enterprises 'lose interest in technical changes that slow up the growth of production and the fulfilment of the year's plan', and Kornai (1959, p. 139) refers to the practice as the 'conflict between today and tomorrow' — which causes enterprises to fail to develop new production techniques and new forms of labour organisation, improve the quality of products, introduce new products, maintain and renew machinery, train apprentices and to retrain workers and technical personnel.

It might be anticipated that when a new product or process had been developed, a centrally-planned economy would exercise a systemic superiority in diffusing the innovation rapidly throughout the economy. Central planners would simply instruct enterprises to produce the new outputs, or use the new process. Empirical evidence suggests that the reverse is true. The percentage of new output produced by a new process, after any given period of time after its initial introduction, tends to be lower in the USSR than in capitalist economies (Amann *et al.*, 1977), while the proportion of old equipment still in use in East European plants is higher than that in West European enterprises (Sik, 1967, pp. 88–9).

This is partially a reflection of the initial stages of extensive growth, when the overriding purpose was to provide an expanding labour force with increased amounts of machinery and equipment, while maintaining output in older plants. As the expansion of the labour force has decelerated, it has become more critical to provide the existing workforce with improved machinery. At this stage the modus operandi of the full-employment constraint, observed by Granick, has hindered the diffusion of technology within existing enterprises.

The preservation, not just of full employment in the aggregate, but of job security in a specific work place (ie, the elimination of frictional unemployment), is seen by many East European economists, as well as Party officials, as an indication of the moral superiority of actually existing socialist economic systems over capitalism. Granick (1975, p. 24) argues that

> it is considered impermissible, except in very rare circumstances, to dismiss workers on grounds other than those of gross incompetence or continued violation of factory discipline. It is considered morally wrong to force workers to change either their existing trade or their place of work because of the abolition of their existing posts.

In practice some frictional unemployment does exist and young workers in particular may experience difficulties in securing their first employment. Labour mobility largely takes place through voluntary job changes or by the reallocation of labour within an enterprise or association. The emphasis in Eastern Europe is that central authorities are responsible for labour reallocation resulting from plant or workshop closures. The problem is largely avoided by maintaining employment in existing jobs unless, or until, a suitable alternative can be found, and changes in the demand for labour tend to result in overmanning.

New processes or products that change the demand for labour are likely to be resisted by those responsible for implementing them, who must themselves take the responsibility for finding alternative employment for workers made redundant. Furthermore the destructive element of Schumpeterian capitalist competitive pressures, whereby existing firms must respond to changes in production processes or the production of new commodities, or go bankrupt, is essentially absent. Such firms can simply continue to produce outputs no longer in demand or at a higher cost than elsewhere.

As a result of the bias against innovation and the diffusion of new technology within existing enterprises, the responsibility for developing new processes and products must rest with central authorities. This gives rise to the separation of research and development units from productive enterprises. Each ministry has its own research institutes, which receive their finance and instructions not from enterprises but from central authorities. In Hungary, enterprises can now contract work to R & D authorities, but even there, R & D is largely divorced from the productive unit (Abonyi, 1981, p. 143). Hanson (1981a) argues that this is consistent with a conception of the process of innovation as one in which new inventions are 'discovered' by specialised research institutes and transferred down the line into practical production. This process, which reflects the method of plan formulation and the political process of democratic centralism, has severe technical (as well as systemic-economic) drawbacks.

In practice, many successful development procedures do not follow this 'linear' sequence. R & D institutes need to be in regular contact with enterprises and production personnel to observe what production problems are and to direct research into appropriate channels — initial development problems and unforeseen developments may require a complete re-examination of the problem at the research stage. There is therefore an additional organisational bias that deters successful innovation.

Extensive and Intensive Growth

Rates of growth of the industrial labour force, industrial capital stock, industrial output and net material product which exceeded world averages, were maintained in each East European country in the first two five-year plan periods. At this stage of development new technology was brought into operation at an appreciable rate by the construction of new enterprises. It was, however, noticeable that the incremental capital output ratio was rising appreciably in each country resulting in a slow-down in the growth of net material product in 1961–5 which was most severe in Czechoslovakia, the GDR and Hungary, where the growth of the industrial labour force was also slowest (Tables 3.1 and 3.2).

The events leading to, and following from, the Polish disturbances in 1956 stimulated official discussion of economic reform in Poland, while the worsening growth performance stimulated an articulate debate on the nature of the economic system in Czechoslovakia and Hungary which culminated in the Czechoslovak reforms of 1967 and 1968 and the launching of the Hungarian New Economic Mechanism in 1968. Unfavourable comparisons with living standards in the Federal Republic led to the idea of reform being discussed in the more conservative GDR, and the publication of the Liberman proposals in the USSR in 1962 made a deep impression throughout Eastern Europe, not so much for their content but because they had been published in *Pravda* (Brus, 1979, p. 267).

The analysis of the Czechoslovak reform movement emphasised Marx's distinction between extensively and intensively expanded reproduction, calling these the stages of 'extensive' and 'intensive' growth. Extensive growth takes place through the expansion of capital and the labour force (increasing input with a relatively slow increase in technique) (Sik, 1967, p. 50). Selucky (1972, Chapter 1) argues that economic development in the first stages of industrialisation in the nineteenth and early-twentieth centuries had been typified by extensive growth but in the post-industrial stage economic growth had depended on a more efficient use of labour and capital (including the application of scientific methods to production and the organisation of production) and improved technical competence of the workforce and management.

Sik (1967, pp. 49–60) proposes that these categories are relative characteristics by which the economic development of different countries at any one point in time may be compared and that the dividing line between them 'will fluctuate during the course of history'.

Sik's definition distinguishes between economies more dependent on innovation (the domestic development of new processes and techniques of production) and diffusion (the spread of existing improvements in technique throughout the economy) as sources of economic growth. These categories are not stages that must be passed through in a fixed sequence but categories that will be more or less appropriate at different stages of development. Presumably, however, a relatively less-developed, but not too backward economy, will derive greater benefits from a policy of extensive growth which permits it to diffuse existing international technology throughout its economy (Gomulka, 1971).

Sik's analysis also gives rise to the possibility of discrete fluctuations in the rate of growth (supply-determined cycles) following a major innovation, or change in production patterns, which ousts old processes and leads to a high rate of growth in labour productivity which will eventually subside, until replaced by a further burst of innovative activity. Marx proposed that such innovations would change the relations of production which would in turn require changes in productive institutions; similarly, Sik argues that the pursuit of policies appropriate to the stage of extensive development, when the appropriate conditions have subsided will have adverse economic effects, and that a policy of extensive growth was appropriate in Czechoslovakia (and therefore Eastern Europe in general) in the immediate post-war period, largely because of the existence of underutilised labour in agriculture. The transfer of labour from agriculture to industry increased productivity in both sectors, as the incremental industrial labour force was equipped with machinery of above-average productivity, and gross agricultural production still grew with a reduced labour force.

The possibilities for expanding the industrial labour force declined drastically. Towards the end of the 1950s, Czech economists started to warn of the prospect of worsening agricultural performance if labour continued to transfer to industry, and by 1963 74.3 per cent of women between the age of 15 to 55 were already employed (Sik, 1967, p. 52).

The continued pursuit of policies geared to replicating existing technology in new plants while maintaining the existence of old plants, combined with the constraint of attempting to maintain existing jobs, frequently resulted in workers with industrial skills and experience continuing to work with ageing and obsolete machinery, while the new plants were undermanned (quantitatively and qualitatively) and their productivity 'did not exceed or even attain [that] of the old factories' (Sik, 1967, p. 60). It was on this basis that the reformers argued that the stage of extensive growth had run its initial course.

Reform Movements

After an initial, unsuccessful appearance in the late 1950s, the idea that some type of reform of the economic mechanism was essential to recreate lost economic dynamism had gained a broad degree of acceptance in the central East European countries by the mid-1960s. There was a far lower degree of consensus amongst both economists and politicians on the precise nature of the reforms that would be either necessary and/or desirable.

The majority of the reform proposals were well grounded in Marxian theory and had in common the need to combine value (price) criteria with elements of central planning, to improve the accuracy of information provided to decision-makers, to avoid over-burdening central planners with excessive inputs of (frequently inaccurate) information and to limit the number of detailed decisions that either had to be taken by Party apparatchiki or were taken on purely political grounds. The various proposals combined these elements in different ways, overlapping in some respects and diverging in others.

The basic theoretical economic arguments can be illustrated by an analysis which combines the methodology developed by Drewnowski (1961) with that developed in the USSR by Novozhilov and analysed by Ellman (1968).

Drewnowski argues that the population has two ways of affecting resource allocation in order to satisfy economic needs; the individual way by earning income and demanding commodities, and the collective way by inducing the state to undertake responsibility for the resource allocations it demands. There arise therefore two forms of preferences — individual preferences and state (planners') preferences. Certain preferences that the individual may have for collective consumption (eg, national defence, freedom from pollution) cannot be expressed by the individual operating in the market and will have to be transferred to the area of state (planners') preferences.

In a democratic society an attempt may be made to establish the wishes of the population through some form of voting system, but it will always be impossible to reveal those preferences in a perfect fashion. As individuals' views will differ in their interpretation of the composition and desirability of state preferences, (legal) compulsion will be required to put them into force. In practice, however, particularly in an undemocratic society, the state preference function may purely reflect the wishes of a dominant class or elite.

The Novozhilov/Ellman analysis proposes that the system of economic organisation involves a combination of three theoretical extremes — perfect centralisation, perfect decentralisation and perfect indirect centralisation. Perfect centralisation requires that all decisions reflect planners' preferences, are actually taken by the planners themselves and are implemented by a series of specific instructions issued by the centre to participants in the economic process (involving, therefore, omniscience and omnipotence). Perfect decentralisation requires that all decisions are taken by individuals (consumers and producers) without instructions from the centre and requires a perfect market or a form of utopian communism to give it effect. Perfect indirect centralisation requires that all decisions are actually taken by individuals, but that those decisions correspond to those of central planners (or those central planners would have taken) because of the criteria specified by the central authorities themselves.

The concept of perfect indirect centralisation can, when combined with different types of planners' preferences, result in widely diverging economic and political systems. At one extreme planners might compute an 'optimal' plan and instead of issuing precise instructions to enterprises issue a set of value parameters and instruct enterprises to pursue a goal such as profit maximisation or sales maximisation. Such a system would be identical with planners' preferences which could reflect the wishes of the population as a whole, or purely those of an elite or undemocratic group.

At the other extreme perfect indirect centralisation is consistent with a situation in which not all information is processed at the centre, but lower-level organisations (enterprises, combines) pursue centrally-determined goals (eg, profit maximisation) on the basis of information received from below (eg, prices in a market). While the concept of 'perfection' is maintained, the resulting resource allocations would be identical to those that *would have been attained* by a group of central planners whose aim was to satisfy consumer demands issuing direct or indirect instructions.

Preferences arrived at without the direct intervention of central planners would be individual preferences, and the system of economic and political organisation would differ greatly from the Stalinist system with a far greater degree of participation in direct decision-making at the enterprise level.

Once the concept of 'perfection' is dropped and imperfections in the process of plan formulation and implementation are admitted, different resource allocations will result from systems of centralisation and indirect centralisation.

In practice, perfect (or total) centralisation would be approximated by a system in which state preferences were supreme and were expressed through a moneyless planned economy embracing full rationing of consumer goods. The operation of the Stalinist planning system in practice involved all three of the categories analysed by Novozhilov. Specific orders and instructions were issued to enterprises, but the problems of aggregation and poor information flows had the effect that decisions taken by enterprises diverged from planners' intentions and were therefore effectively decentralised, but still did not reflect individual consumers' preferences. Where aggregated indices did have the desired effect this was more consistent with the idea of indirect centralisation consonant with planners' preferences.

There was therefore considerable scope for reform and (apparent) decentralisation that actually strengthened the powers of the centre and which, by improving the flow of information from enterprises to planners and the system of instructions, would encourage enterprises to act in a way that would coincide with planners' intentions. In certain cases improved prices (including those derived from markets) and profitability criteria were perfectly consistent with improved centralisation.

Drewnowski (1961) has described the systems that operate within a framework in which state preferences are dominant but individual preferences play an increasing role as first-, second- and third-degree market economies. The Stalinist system is a first-degree market economy in which the state determines the output of consumer goods in a considerable degree of detail, but their distribution between individuals is left to the operation of (imperfect) labour and retail markets. In the second-degree market economy the determination of the quantities and types of output from existing plant is left to the market. In the third-degree market economy investments in new plant for consumer goods are also determined by the market (subject to the macroeconomic constraints on the size of total consumption and on total investment in the consumer goods sector imposed by the state).

Neuberger (1966) and Keren (1973) have divided the actual reform movements in the USSR and Eastern Europe into three categories: *centralised reformers* who wished to maintain the power of the administrative hierarchy to determine large decisions while leaving the smaller decisions to be determined by enterprises, but who proposed devising methods whereby the sum of the small decisions would be equal to the decisions taken at the centre ('Liberman and his like'); the *computopians* who wished to improve the ability of the hierarchy to process information by the use of mathematical methods; and the *market socialists* who believed that no amount of improvement to the existing planning system would correct the fundamental ills of the economy and wished to transfer economic decisions to autonomous enterprises responding to market demands.

These categories approximately coincide with Lewin's (1975, p. 186) distinction between the 'supporters of administrative methods' and the 'supporters of economic methods' and the division of the second category by Selucky (1972, p. 43) (who helped design the economic and political programmes of the Czechoslovak reform movement) into supporters of 'technocratisation' and 'marketisation'. Lewin calls the supporters of administrative methods 'statists', as they resist attempts to weaken the role of the state bureaucracy, although the term may be slightly misleading as a major concern of the statist mode of production is to preserve central Party control over the economy (eg, Ceausescu's Romania) but can be justified by the view that the Party itself has become statised.

Selucky (1972, pp. 43–55) argues that supporters of technocratisation do not wish to change the fundamental nature of the command system, but wish to modify it and transfer political control from the Party apparatus to the 'technical and economic experts who pursue purely pragmatic rather than ideological aims' (ie, a managerial elite will replace the Party elite at all stages of decision-making). For Selucky (1972, p. 18) the important distinction is to be made between marketisation and non-marketisation — 'the entire mechanism of the non-market hierarchically centralised model can only stimulate extensive not intensive growth'.

Selucky's economic and political programme can be summarised as follows: the hierarchical system of command planning will be replaced by indicative planning, centralised decisions will be decentralised and 'marketisation [will form] the basis of an overall social reform of the Soviet-type system' (1972, p. 43). Autonomous economic units will respond to market prices which will provide objective data on needs (demand) and costs which will also form the basis of rational investment decisions. Enterprises will operate on the principles of self-administration (workers' control — which will allow the working class to participate directly in decision making and become the ruling class) and self-interest (ie, the pursuit of profits with a dividend to be distributed amongst the workers of the enterprise). The replacement of the vertical chain of economic command by horizontal contacts between enterprises will help to break down the vertical political chain of command and lead to its replacement by a horizontal pluralist system.

Brus' Programme: The Regulated Market Economy

Selucky's analysis of the political intentions of those whom he terms 'supporters of technocratisation' is a harsh judgement on many members of the mathematical-cybernetic school of reform, whose models were frequently used to demonstrate the shortcomings of the centralised system of instructions and the advantages to be gained from (and the legitimacy of) the utilisation of value (price) criteria, and who were often in the vanguard of the movement for decentralisation of the economic system and some form of democratisation of the political system. Brus (1975, p. 195), for example, denies that the improvement of planning techniques will either require, or lead to, the emergence of a technocratic elite — 'the large and ever-growing role of experts does not conflict with the possibility of or the need for the democratic form of choice'.

Brus' economic proposals for a 'regulated market economy' were first published in Poland in 1961, and his political ideas have been published subsequently in the UK with the result that his collected ideas form the most coherent economic and political programme of the 'indirectly centralised' type. A critical feature of Brus' argument is that the stifling of innovation by lower-level economic units results from the political system of Stalinism not from the economic system of central planning.

Brus' principal arguments are:

(1) Public ownership (nationalisation) of the means of production has resulted in the statisation of the means of production, not genuine social ownership which will only occur if the political system ensures the subordination of the socialist state to the wishes of the people. In the Stalinist political and economic system the Communist Party has become the instrument by which the ruling

group has extended control over all aspects of economic and social life. State preferences in the economic field represent the preferences of a bureaucratic elite, not the wishes of the people as a whole.

(2) Workers' control of the Yugoslavian type alone (or its theoretical Illyrian counterpart) will not ensure true social ownership as workers will only control a subset of the economy (the enterprise). In administering the enterprise solely on their own behalf workers will take insufficient account of the effects of their actions on others (ignore externalities). In particular the pursuit of the Illyrian goal of maximum income per worker within a given enterprise will not ensure full employment, or maximum output with available resources, or a socially optimum level and structure of investment and it will reinforce income inequality between enterprises. The operation of any (pure) market system ignores broadly-conceived social criteria and is socially sub-optimal.

(3) The state must act to secure socially desirable goals in the economic sphere through the imposition of state preferences. The most important aims of government economic policy are to establish macroeconomic goals corresponding to social preferences, and in particular to ensure full employment. This involves establishing and securing investment ratios, the division of investment into productive and non-productive (social) investment, the nomination of certain specific investment projects, the conduct of social policies and policies towards income distribution. In addition, however, apparently technical policies such as the protection of the environment and the degree of automation must be determined by state preferences reflecting the wishes of the people as a whole, not market preferences which will only reflect the individual's preferences in his role as a consumer (eg, cheap nuclear energy).

(4) State preferences will only reflect people's preferences if the system is truly democratised. The political system that will achieve this is not fully elaborated but could include 'parliamentarism, linked with worker, territorial and co-operative self-management, with independent trade unions, etc' (Brus, 1975, p. 197). A critical feature would be the positive selection of personnel at all levels by democratic processes, not bureaucratic fiat. The existing totalitarian *political* system acts as a constraint to economic development by rewarding docility and servility, by hindering the development of new ideas, and by alienating the working class in general which increases resistance to change and lowers productivity.

(5) The long-term continuity of effort and commitment, the inclination to take initiatives and to perform 'good work' will only be guaranteed if the workers' individual interests and those of society are harmonised. This will require not just the payment of monetary incentives, but genuine feelings of participation and social responsibility.

(6) Central government economic policies will not be determined by physical instructions to enterprises but by central controls over prices including those for labour (wages) and capital (interest charges) which will be issued to enterprises and will form the basis of their decisions on the choice of inputs and outputs. The state will adjust prices to make enterprise decisions correspond to state preferences, while the payment of individual and collective material incentives based on enterprise results will provide a stimulus to productive activity, provide workers with an incentive to increase labour productivity, and reduce the consumption of raw materials.[1]

Czechoslovak Reform Proposals

The Czechoslovak economic reforms proposed to give a far greater role to market demands in the determination of short-term decisions than that envisaged by Brus' model. The essential economic distinction between the Czech reform proposals and those of Brus can be summarised by the following statement from the Deputy Chairman of the Czechoslovak State Planning Commission (Kohoutec, 1968, p. 126): 'those concepts which regard the market as some instrument for carrying out the plan are unrealistic since the market cannot be considered a category subordinated to the plan'.

The economic system that would have emerged if the reforms had been fully implemented was summarised by Holesovsky (1977, p. 714) as 'market-type state capitalism, guided by econometric forecasting, tools of indicative planning, opened up to foreign competitive imports, possibly with elements of employee participation in management, and surely with Trade Unions returned to their role as autonomous interest organisations of labour'.

This would be a guided market economy in which the market played a major role in determining short-term output decisions, but the state's long-term objectives would be implemented by credit policy and shorter-term macroeconomic equilibrium would be determined by wage and price policies. The banking system would play a role similar to that envisaged by Hilferding and Lenin, more interventionist than that typically played in a western market economy and more responsive to state plans, but monetary criteria would play a more active role than was typical in the Stalinist system.

Central planning was to be confined to developing medium- and long-term macroeconomic targets, and annual central plans would no longer be drawn up. The macroeconomic targets would not be disaggregated amongst enterprises in the form of binding plan instructions.

Enterprises would not receive compulsory plan indicators but would be motivated to maximise gross income, defined as net value added (including labour costs) or as total revenue minus the costs of materials and depreciation (Sik, 1967, pp. 233–7). Gross income was then to be divided between payments for basic wages and bonuses, funds for social and investment activities, funds for reinvestment and technical development and tax payments, and loan repayments. The state would influence enterprise activity and the distribution of funds, not through binding instructions but by fiscal policy, whereby tax payments would be differentiated according to the uses to which funds were put.

Enterprises would have far greater autonomy, not just in their choices of inputs, outputs and suppliers, but in their freedom to join and leave associations and decide their own organisational structure (Kyn, 1975, p. 114). The majority of these decisions, including the appointment of managers, were to be decided by workers' councils, elected by the workers themselves. The development of medium- and small-scale enterprises was to be encouraged, including the foundation of new enterprises, which could be established by existing enterprises or other authorities. After such establishment, the new enterprises would become independent self-managed organisations.

In the longer term it was anticipated that retail prices would be determined by market forces, but in the short term would be subject to ceilings and some central price-fixing to prevent inflationary pressures. Central government controls over prices and wages were only to remain for macroeconomic reasons and to regulate monopolies.

This model diverged considerably from Brus' model in that prices and enterprise outputs were to be determined by individual preferences expressed in the marketplace not by the state acting on behalf of individuals and issuing prices to enterprises. The latter was considered unrealistic by Czechoslovak mathematical economists because of the inadequacies of data supplied to the centre and the limitations imposed by computer size on the number of prices that could be estimated centrally. Computerised calculation of the long-term market price trends that would accompany centrally-determined policies would serve as a guide to central planners to shape the tax policies to harmonise enterprise and central interests — 'the planned price should be an equilibrium market price, with current market price moving according to momentary supply and demand' (Kyn *et al.*, 1967, pp. 102–3).

Brus and the Czechoslovak Reform School: A Comparison

The proposals of Brus and the Czechoslovak market reform movement have a number of features in common. Both analyses propose (i) the unreformed Stalinist system was inappropriate for the stage of development reached by the central East European economies by the early 1960s; (ii) that the state and Party bureaucracy do not operate in the interests of the population as a whole leading to increased alienation, the economic effects of which are reflected in reduced labour productivity; (iii) material incentives linked to success indicators (especially output) can have short-term effects on productivity but these are frequently socially irrational; (iv) a bureaucratic hierarchy, particularly one in which those who occupy senior positions claim the unchallenged right to interpret what is morally beneficial for the people as a whole, stifles initiative and creates an environment in which those who contradict central decisions are penalised.

Both analyses propose that the problems of the bureaucratic economy can only be overcome by radical reforms to the centralised system of vertical planning, including the abolition of the material technical supply system and its replacement by horizontal links between enterprises on the basis of contracts determined by the enterprises themselves, the replacement of the cumbersome system of detailed instructions to enterprises by a single maximand (profit, gross income) derived from rational market-based prices, the participation of workers in all stages of the decision-making process, the linking of workers' material interests to social interests through socially rational indicators of enterprise performance, and the transformation of trade unions into genuine workers' organisations.

The crucial difference between the two schools is that Brus argues that it is alienation induced by the *political* system which both requires central authorities to issue detailed instructions linked to material incentives, and provides enterprises and workers with the possibility and incentive to manipulate those instructions to achieve personal rather than social objectives. Brus proposes that the reform and democratisation of the political system will remove this alienation and lead to improved economic performance. Furthermore, because the centre alone is capable of interpreting issues from a social perspective, this economic performance will be better than that provided by a socialist or capitalist market system. The role of the planning system should therefore be to harmonise enterprise and workers' interests and devise a rational signalling system corresponding to those interests; central planning itself is not to blame for the economic ills of Eastern Europe.

Ota Sik and the Czechoslovak market reformers argue that the replacement of central planning by a market system is a prerequisite to the breakdown of the political system: 'an economic structure once developed further determines the character of political changes in a later period' (Sik, 1967). Even if central planning could be made to work more efficiently, the problems of the bureaucratic economy would inevitably re-emerge to hinder innovation.

The question of which system would provide a greater stimulus to product and process innovation and diffusion remains unanswered. It is a logical inference from Hirschman that the incorporation of the social costs of innovation into central decisions will reduce the incentive of the state to innovate and in particular to diffuse new technology.

Brus argues (1975, pp. 201–2) that the new socialist entrepreneur will be motivated to avoid over- or under-innovation by his self-identification with the needs of society, and presumably therefore new innovations would be diffused at a socially optimal rate as old plant became redundant. The gradual incorporation of technical improvements under conditions of social harmony should be seen as a desirable end in itself.

Schumpeter (1974, p. 83) viewed this as a final ideal goal when creative destruction stimulated by individual self-interest is no longer necessary, but warns that 'any system that at every given point in time fully utilises its possibilities to the best advantage may yet in the long run be inferior to a system that does so at no given point in time'.

Schumpeter's criticism can be levelled at any market-type proposals based on short-term static welfare maximisation. In the East European context these criticisms would be less important if the economies concerned possessed no superpower illusions, and were not dependent to any substantial degree on trade with capitalist economies where Schumpeterian competitive dynamics will apply.

Similarly, therefore, the method by which a sufficient incentive to innovate will be retained while simultaneously providing social justice for those who have been 'innovated against' is not entirely apparent in the Czechoslovak market proposals.

The Implementation of Reforms in Eastern Europe and the Barriers to Reform

A series of reform proposals were first discussed in the USSR, Poland, Hungary and Czechoslovakia and to a lesser extent the GDR in the mid-1950s. These proposals could be described as 'indirectly centralised' and were intended to improve the performance of the existing economic mechanism by providing incentives and signals to enterprises that would bring a greater correspondence between their interests and those of the central planners. The proposals included a reduction in the number of compulsory indicators and a corresponding decentralisation of microeconomic decisions to enterprises, the replacement of the vertical chain of command by improved horizontal links between enterprises, linking enterprise and workers' incentives to profitability, and in certain cases, especially Poland, establishing workers' committees in enterprises (Brus, 1975, pp. 149–50). The implementation of these proposals was either slowed down or abandoned after the events in Poland and Hungary in 1956, although movements towards reform continued in Czechoslovakia in 1958 and workers' councils were established in Polish enterprises.

A significant *political* event in reform history was the publication in *Pravda* of Liberman's indirectly centralist reform proposals on 9 September 1962.[2] Major economic decisions concerning the determination of macroeconomic aggregates, large capital investments, financing and prices were to be determined centrally, but enterprises were to have greater freedom to make detailed microeconomic decisions within their sphere of competence. Enterprises would receive aggregated output and assortment plans and would draw up production schedules in consultation with central authorities, specifying the labour inputs, new investment, new technology, and so on, required to meet these targets.

Managers' and workers' interests were to be harmonised with central interests by material incentives linked to a profit rate calculated on the basis of revenue received minus operating costs (including wages) expressed as a percentage of fixed and working capital. The introduction of costs into the calculation of bonuses provided enterprises with an incentive to seek the least costly methods of production and to reject sub-standard inputs which would enter into costs, while revenue received would mean that enterprises would not be rewarded for the production of rejected outputs.

Enterprises were to be encouraged to adopt taut plans and provide accurate information to the centre by a procedure whereby the higher the profit rate the enterprise *planned* to achieve the higher the rate it would be *deemed* to have achieved, if fulfilled. The payment of bonuses was to be conditional on the fulfilment of gross output and assortment targets. Profits were not to be retained by enterprises, but were to be both a success indicator and a source of finance for bonuses established on a predetermined basis with remaining profits accruing to the state budget.

In the GDR, Liberman-type proposals were santioned as a reform mechanism in 1963, but the reforms were effectively dismantled in 1969 (Keren, 1973). In Czechoslovakia reform discussions were initiated in 1964 and the resulting economic reform initiated in 1967 involved elements of marketisation and workers' control as shown above, but was gradually dismantled from 1969 onwards, following the (1968) Warsaw Pact Invasion. Kosygin's October 1965 reform proposals for the USSR were far more centralist than the initial Liberman proposals. In Bulgaria, Liberman-type reforms (but incorporating a greater degree of price flexibility and decentralised investment) were approved in December 1965 but the new system was largely dismantled from July 1968 onwards (Vogel, 1975). Ceausescu initiated reform discussions in 1965 in Romania, and the reform directives approved in December 1967 were mainly of the Liberman type, but also implied the possibility of a degree of marketisation combined with the principle of collective management. The radical elements of these proposals were shelved in 1968 (according to an emigré source) following the events in Czechoslovakia, but have been formally reopened, though inadequately implemented, since 1978.

The degree of sophistication of Polish reform proposals has not been matched by a corresponding willingness to implement them by the Polish authorities, resulting in an ebb and flow of the policies adopted. Only Hungary has successfully initiated reforms of the 'regulated market' type and sustained the principle of the reform, although some recentralisation has occurred.

Western analysts of the East European economies have frequently been unable to ascribe the retardation or reversal of reforms to domestic economic factors but have sought political explanations, particularly as the first wave of reforms was abandoned after the Polish and Hungarian uprisings of 1956 and the second series was (with the exception of Hungary) abandoned at the same time as, or following, the Warsaw Pact invasion of Czechoslovakia.

It is difficult in these circumstances to distinguish between Soviet opposition to reforms, domestic opposition and effective 'self-censorship' in anticipation of Soviet opposition. Brus (1981, p. 85) has argued that it is unlikely that the USSR would seek to block reforms in Eastern Europe that were restricted purely to the operation of the economic system and that reforms of the regulated market kind are acceptable provided they are not conceived as part of a 'broader political blueprint aiming at . . . democratic pluralism connected with national aspirations and the denunciation of Soviet-type totalitarianism', which would result in conflict with both the Soviet leadership and domestic conservative factions.

It is difficult for a leadership to decide whether purely economic changes will stimulate demands for political change. The Czechoslovak reform proposals clearly intended to use economic change to stimulate political change, while the Hungarian leadership avoided this trap. The partial implementation and lack of success of Polish reforms may well have resulted from the fact that the leadership was only willing to implement the economic aspects of reform proposals while the political proposals which were considered a prerequisite for their success were not implemented. The ensuing economic dissatisfaction, combined with those aspects of an open society which were not tolerated elsewhere in CMEA, but which the Polish leadership was unable or unwilling to suppress, subsequently stimulated the open demand for economic change and the emergence of Solidarity.

Although Liberman's proposals appeared to be directed at improving the operation of the existing system by encouraging enterprises to produce centrally-determined outputs at minimum cost, within the framework of the material-technical supply system, the logical implications of his proposals could require substantial changes to the planning structure.

The choice of least-cost solutions involves the substitution of one set of inputs (eg, steel) by another (eg, plastics), investment in new (possibly labour-saving) capital and technology and therefore has considerable implications for the enterprises's relationships with other enterprises, central authorities and the Party.

A proposed change in an enterprise's input and investment structure requires an initial change in the output structure of supplying enterprises and will result in corresponding changes throughout the economy. If these changes are to be taken, as Liberman suggests, in consultation with central authorities, the initial decision to change inputs must be channelled back to central planners who must recalculate the plan and issue changed output instructions to enterprises. Logically, where central planners are empowered to issue output instructions only, this process would be repeated ad infinitum on the basis of an initial change, although in practice the build-up and reduction of inventories would act as a buffer. The problems involved however could result in considerable delay and inertia and, in particular, the problems of encouraging enterprises to produce new outputs and adopt new production techniques using inputs for which adequate stocks do not exist would not be overcome.

A second problem concerns the prices to be used in enterprise accounts and inter-enterprise transfers. If enterprises are to be motivated to seek and adopt least-cost solutions, prices must reflect the real cost of inputs, calculated on the basis of their contribution to centrally-determined objectives (as would, for example, result from the dual of a linear-programming solution). This would involve a practical problem of calculation and if prices are fixed for long periods (as is the current practice) while supply conditions change, enterprises will make decisions based on incorrect information. Furthermore, if prices are fixed centrally enterprises may be unwilling to accept inputs or produce outputs which offer high social (total) profits, if the gains are unevenly distributed between enterprises. For example, the use of synthetic materials may substantially reduce the cost of producing an enterprise's output but modify its specifications. If prices are fixed, and the cost reduction cannot be passed on to receiving enterprises, the latter will have no incentive to accept the new input. Effectively, the benefits arising from cost-reducing innovations would be 'over-internalised'.

The retention of central determination of enterprise outputs and inter-enterprise prices would still result in a sluggish response to changing circumstances and could discourage enterprises from innovating. The problems point therefore to the establishment of direct contacts (contractual links) between enterprises which permit them to change their outputs in response to changed orders (without central intervention) and to negotiate mutually advantageous inter-enterprise prices, as was contained in the Bulgarian and Hungarian proposals.

The problems of this approach are that wholesale prices would logically be determined by demand conditions in retail markets and in the case of the East European countries where imported raw materials are significant, by conditions prevailing in world markets. The economic effects could conflict with central objectives concerning price stability, full employment and income distribution.

The essential feature of the unreformed centrally-planned economy is that central planners exercise detailed control over inter-enterprise transfers. A system based on direct contacts between enterprises determined by market criteria need not present a direct threat to Party political control, but it will complicate the pursuit of many of the Party's economic objectives that retain a considerable degree of workers' support, and this will, in the eyes of some, pose an ideological threat and also threaten to disperse, or even break up, the industrial ministry system and the system of material-technical supplies.

The factions in both the Soviet and East European leaderships that feel themselves to be most threatened by marketisation proposals are likely to form an alliance to resist such proposals. It appears that in each country except Hungary this coalition managed to limit economic reforms to measures designed to improve the performance of the existing centralised system following the events of 1968.

A major feature of the centralist reforms was the devolution of decision-making to a middle-tier of management composed of amalgamations of enterprises and known by their acronyms as DSO (Bulgaria), VHJ (Czechoslovakia), VVB (GDR), WOG (Poland) and as Industrial Centrals in Romania. In Hungary the 'indirectly centralised' reforms were preceded by substantial enterprise amalgamations, although the term 'enterprise' was retained for the larger units.

The new units spanned a broad range of organisational types (vertical and horizontal integration, etc) within each country and the powers devolved to them varied from country to country and from time to time, with Romania normally the most centralised. The new units were often staffed by former ministry officials but were not intended to become a mere additional link in the organisational chain but were to become genuine production agencies, whose administrative bureaux were frequently located at an enterprise.

The new units generally had in common the fact that they operated on independent accounting (*khozraschet*), received plan instructions that contained less specific detail than those previously given to enterprises, and played a greater role in drawing up their own plans. In many cases, research and development organisations were incorporated into the new bodies to bring R & D closer to production. Input costs were introduced into the method of calculating bonuses by replacing gross output targets by indicators linked to value added (net production) or some profit criterion (although in many cases sales, or change in sales, were also introduced as success indicators) (Jeffries, 1981, *passim*).

Both economic and administrative logic can be found in these moves. Central planners can maintain overall control over inter-enterprise transfers without being burdened with a mass of detail if microeconomic decisions are taken within large production units. Furthermore, if production decisions are to be based on administrative and not price criteria, decision-makers must have a wide view of the implications of their decisions and have the power to implement them over a wide area. Taut planning also meant that the economies operated with substantial deficiencies and shortages of material, labour, and so on, between economic organisations, but with reserves and excess capacity within economic units (due, for example, to hoarding of material and labour). Amalgamating economic units enables these reserves to be redistributed within the new organisation.

This probably accounts for the initial emphasis on horizontal integration (mergers of enterprises producing similar outputs) under the jurisdiction of a single ministry. While this preserved the existing power structure, it also helped to circumvent the 'no dismissals' rule by enabling the transfer of labour between constituent enterprises, resulting in the closure of uneconomic or outdated plant.

Vertical integration across ministerial boundaries has been more noticeable in the GDR, and to a lesser extent Poland and Czechoslovakia and has been principally directed at improving and speeding up inter-enterprise transfers, reducing the incentive to hoard stocks of supplies and making supplying enterprises more responsive to changed demands. Critically, however, the profits arising from a new innovation in a supplying enterprise or receiving enterprise within a combine will now accrue to the combine as a whole, not just the innovating enterprise, and combine management will have an incentive to instruct constituent enterprises to co-operate in the development of innovations occurring upstream or downstream from the enterprise itself.

Hungary alone has enacted a reform of the regulated market kind (Hare *et al.*, 1981). Central policy is principally directed at controlling the macroeconomic aggregates of consumption and investment and at maintaining full employment, balance of payments equilibrium and controlling inflation. Enterprises no longer receive annual plans specifying quantitative output targets and sources of inputs but attempt to achieve profit targets and are allowed to make direct inter-enterprise contacts. Enterprises are guided towards the fulfilment of central objectives by a combination of differentiated tax, subsidy and credit policies. Prices are intended to reflect world market conditions as much as possible.

In practice, central controls remain in many critical areas. The state maintains the power to found new enterprises, to close or merge existing enterprises and to delineate their sphere of production. There can be no spontaneous development of new products, or new entrants to the market or product diversification by existing producers without central approval. The high degree of geographic concentration of industry around Budapest also means that enterprise managers are in close contact with central authorities and may receive spoken instructions where no written ones exist.

Enterprise amalgamations have strengthened enterprise monopolies and the desire to maintain full employment has resulted in the state maintaining a high degree of aggregate demand and enterprises are frequently motivated to maintain a growth of output. A sellers' market still persists therefore, and frequent shortages occur. Enterprises find it difficult to change suppliers and, where enterprises have been empowered to negotiate prices subject to centrally-determined ceilings, prices frequently rise to those ceilings and are *de facto* centrally fixed.

Central authorities were forced to reinstitute central controls to preserve income equality from 1970 to 1972 and to enter into enterprise-specific subsidies to prevent inflation and windfall income gains and losses, following the increase in world oil prices. (A more detailed examination of these effects is given in Chapters 7 and 9.) Consequently, enterprise managers may still achieve more for themselves and their workforce through the process of bargaining with central authorities than through generating economic improvements.

Despite these criticisms, it is true, as Radice (1981b) argues that Hungary has still managed to maintain the *principle* of the reform intact and is currently engaged on reaffirming that principle.

Corporatism?

A major feature of the post-1965 reform era has been the growth of enterprise size. In Romania, in addition to the establishment of industrial centrals (112 of which incorporated 1,635 enterprises and 2.7 million workers in 1977), enterprises themselves have undergone substantial reorganisation and amalgamation so that by 1977 85 per cent of industrial employment and output were provided by enterprises with more than 1,000 blue collar workers (Granick, 1975, observed that the enterprise is more of a workshop or production unit). Between 1972 and 1977 the entire growth of industrial employment (663,000 workers) was absorbed by enterprises with more than 3,000 workers (Smith, 1981b). In Czechoslovakia and Hungary over 75 per cent of output is provided by enterprises employing more than 1,000 workers, while in Bulgaria the average employment in a DSO is 17,000 workers (Kaser, 1981). In the GDR, combines have swallowed up many smaller, more market-oriented producers.

The degree of enterprise concentration is far larger than in any western country and has been accompanied by a growing social role for the enterprise. In Romania many of the former functions of local authorities are being provided and financed by enterprises, including employee housing and even some aspects of secondary schooling. The enterprise now plays a substantial role as the tax gathering unit in Romania and Hungary and the right to participate in certain aspects of the social welfare system is restricted to state employees.

There is therefore evidence of a tendency towards corporatism, particularly in Romania. In addition however the unresponsiveness of large units to providing after-sales service and spare parts, may have helped to stimulate the development of unofficial private market activities.

Notes

1. See in particular Brus (1972, 1975) and Gomulka (1977).
2. An English translation appears in Nove and Nuti (eds) (1972).

PART TWO

WAGES, RETAIL TRADE AND CONSUMER EQUILIBRIUM

5 MONEY, BANKING AND CONSUMER EQUILIBRIUM

The Role of Money in the Stalinist Economic System

> Communist Society will know nothing of money. Every worker will produce goods for the general welfare. He will not receive any certificate to the effect that he has delivered the product to society. He will receive no money . . . in like manner he will pay no money to society when he receives whatever he requires from the common store . . . A very different state of affairs prevails under Socialism.
>
> Bukharin and Preobrazhensky, *ABC of Communism*

The quotation above typifies the distinction made by Marxist thinkers between the role played by money under communism and under socialism, and its relationship to the system of distribution. Under communism, goods and services will be distributed according to criteria of need, rather than according to the amount of work performed, and the worker will not require payment in the form of money income, while work itself will be performed more as a social duty than for purposes of individual gain.

Although Marx considered that this would require such an abundance of commodities that scarcity and therefore choice would become irrelevant, it can be argued that allocation according to need does not necessarily imply that each individual should be free to consume whatever he wishes, but that need may be determined by some other body, according to social criteria (eg, family size) and/or objective criteria (eg, medical or educational need).

Marx and his followers argued that money would still have some role to play in a socialist economy, when allocation would be according to work performed. Marx argued that under socialism workers would receive vouchers entitling them to withdraw consumer goods equivalent to, or proportional to, the amount of labour time provided from a stock of social supplies. As these vouchers did not circulate he argued that they were not money. Preobrazhensky refined Marx's analysis and argued that a major proportion of wage-income should be allocated directly at, or by, the workplace in the form of canteen meals, housing, health facilities, education, and so on. Only a small proportion of income would have to be distributed in the form of vouchers (Day, 1975). If these are not considered to be money the abolition of money could become a theoretical possibility even under socialism, while the transition to allocation according to need by the workplace would also be simplified.

Preobrazhensky (Bukharin and Preobrazhensky, 1970 edition, p. 90) envisaged that if the state sector were to coexist with a private sector, money would still be necessary for the role it would play in transactions between the sectors. Certain sections of the Soviet leadership still entertained the idea of a moneyless economy during the first years of the five-year plan, but this concept was finally abandoned and a system that retained individual enterprise accounting, the payment of money wages and the continuance of private trade (particularly for agricultural commodities) was accepted as providing the basis of a socialist economy (Lewin, 1975, pp. 97–124).

The process of maintaining macroeconomic equilibrium continued to be refined throughout the 1930s. The budget became the central mechanism for collecting funds and financing investment. Between 1931 and 1935, 80 per cent of capital investment came from the budget while turnover taxes (effectively the difference between the cost of production and the state selling price of a commodity), and deductions from enterprise profits became the major source of budget revenue (Podolski, 1973, pp. 32–57).

The banking system continued to play a considerable role in the disbursement of credit to enterprises. The provision of credit for working capital became the monopoly of the State Bank which had to ensure that enterprises produced according to plan before sanctioning payments for inputs and wages. Credit for capital investment was supervised by specialised banks, which remained part of the state banking system. Consequently the system of credit provision strongly resembled the Real Bills Theory of Adam Smith (of which Marx was an adherent) in that monetary policy was basically concerned with the supply of credit to productive industry, rather than the control of money supply in aggregate. Money supply is therefore endogenous to the economic system and follows other economic variables rather than determining them (Podolski, 1973, pp. 12–17).

Provided credit is only advanced against genuine 'work in

progress' this is not held to be inflationary in the long term, as output will be provided to match monetary advances. Podolski (1973, p. 53) argues that credit provision in the early 1930s was based on a very loose interpretation of this principle, and that credit was frequently advanced to pay for above-plan labour inputs and material consumption with no additional output resulting from their use. Consequently, the money supply grew five-fold between 1927 and 1934, while the industrial labour force grew 2.3 times and the average industrial wage 2.5 times. Significantly, however, this overpayment of wages was not allowed to divert resources from investment to consumption, due to the passive nature of money. The system of bank administration of enterprises was modified in 1931 and 1935, but the most significant modification that gives the Soviet and East European monetary system its most distinctive feature — the creation of two separate monetary circuits (involving cashless inter-enterprise transfers) was made in 1939. As a result, cash is only injected into the system with the payment of wages which can then be utilised for the purchase of consumer goods and services. Transfers of material inputs between enterprises follow plan instructions and enterprise accounts are debited and credited accordingly by the State Bank without any actual cash transfers taking place.

The 'Passive' Nature of Money in a Centrally-Planned Economy

A major distinction between the function of money in a centrally-planned economy and its function in a market economy is that money in the former is frequently described as being 'passive'. A simple definition of passive money is that the possession of money alone does not enable the owner to acquire resources or goods and services whose allocation has been determined by a plan instruction. Passive money therefore functions purely as a unit of account, and money flows follow, rather than determine, resource flows.

In practice, the degree of passivity of money varies considerably in different sectors of the economy, and the term 'passive money' applies most accurately to inter-enterprise transfers of material inputs, the centralised allocation of investment supplies, and so on, which are matched by cashless debits and credits and can only legally take place after a corresponding allocation certificate has been issued.

The principal effect of passive money in this sector is that the enterprise cannot purchase inputs other than those established in the plan, cannot change suppliers from those established in the plan, and cannot (legally) use money balances to acquire inputs in short supply. Where monetary transactions are used to acquire unspecified inputs they are more likely to take the form of illegal cash payments to individuals rather than bona fide inter-enterprise transfers. A major systemic factor that arises from the passivity of money in this sector is that the enterprise has no financial incentive to dispose of physical assets which are surplus to its requirements (or which could be put to better use elsewhere) or to diversify its production on financial grounds.

A principal feature of cash payments is their anonymity. Cash payments cannot easily be traced, making cash an unsuitable medium for passive money. In markets where cash payments are allowed therefore money tends to be more active, although the degree of activity varies considerably. Enterprise money is most active in the labour market where it gives the enterprise command over productive resources (ie, labour), but even here the degree of activity of money is limited by the constraints of the labour plan, which may be violated, but not excessively.

Money has various degrees of activity in the market for consumer goods and services. Some major items which are basically consumed privately (eg, housing, education, health facilities and even on certain occasions recreational facilities) are subject to direct allocation for which a corresponding passive monetary payment will have to be made (a position analogous to the provision of local authority housing and university education in the UK).

Money income can be spent by the individual in state retail stores and in various legal and illegal markets. The use of money in state retail stores bears a strong resemblance to the voucher system espoused by Marx, in that consumers can choose between the set of commodities that planners make available to them at fixed prices. Money is active in this sector only in the sense that it allows the consumer to choose between the commodities that have been made available, but in all other respects is passive, as planners need not adjust either the total quantity, the commodity composition or even the prices of goods made available in response to monetary demand. As a result, any overpayment of wages will not divert resources away from planned investment and central expenditure.

Money is more active in the legal and illegal private markets than in the state retail markets, to the extent that suppliers' decisions concerning what to produce and where to sell their produce will be affected by monetary demand. The activity of money may not extend however to the supply of factors of production (other than labour) and intermediate goods to these activities. Consequently, private producers may be unable to acquire basic inputs (eg, agricultural machinery) unless the resources required for production of these commodities have been allocated by a plan instruction. This effect is reinforced by the foreign trade monopoly which prevents individuals from acquiring inputs from abroad.

Microeconomic Equilibrium in the Labour and Consumer Markets

The Supply and Demand for Labour and Sources of Money Income

The role of the industrial employee in obtaining and spending money income in the East European economies has many initial similarities to that of his counterpart in the state and corporate sectors of the mixed economies of Western Europe.

Enterprises in Eastern Europe receive a labour plan which currently varies from a detailed plan specifying the number of white-collar and blue-collar workers to be employed and the wage rates to be paid to specific labour categories (Romania) to a constraint on the total wage bill (Hungary).

Workers are free to choose their place of employment within certain limits. Constraints on the free mobility of labour have varied through time from the use of forced labour to the system of work passports involving the need to acquire permission to live and work in major cities (Granick, 1975, p. 68). Certain categories of the population (notably graduating students and young Party members) may be directed to specific places of employment. In Hungary, a measure that is expressly designed to direct services to provincial areas requires newly-qualified doctors and teachers to choose a post from a central list and spend a minimum of three years there (Gabor and Galasi, 1981).

A major preoccupation of government policies in the labour market has been the avoidance of unemployment, arising either from deficit demand in aggregate, or from changes in the structure of demand for specific categories of worker. Consequently, enterprises frequently face constraints in laying off labour, particularly in areas where few alternative opportunities exist, while other enterprises may face difficulties in attracting labour. In an attempt to overcome this problem even Hungary currently resorts to non-market methods. Workers leaving jobs can only take up employment at certain nominated enterprises, while less-efficient enterprises are forbidden to replace workers who leave (Gabor and Galasi, 1981). In Romania enterprise amalgamations have empowered enterprise managers to transfer labour from one workplace to another within the new larger enterprise (Smith, 1981b).

In general, however, the worker can sell his labour to the highest bidder while the enterprise must use monetary incentives to attract and retain labour. As a result even centrally-determined wage rates will be adjusted to meet market conditions — the norms on which piece-rates are assessed may be adjusted to attract labour, labour in short supply will be artificially upgraded, and so on. In Hungary, where central constraints were imposed on the employment of white-collar labour in 1977, clerical labour was reclassified by enterprises as manual labour, causing the measure to be abandoned in 1979 (Gabor and Galasi, 1981).

In addition to income from employment, households receive other forms of state-provided monetary payments in the form of pensions, maternity and child benefits, temporary disability benefits and permanent disability pensions, sickness pay, student grants, and so on.

State incomes may be supplemented by a variety of legal and illegal market activities ranging from moonlighting (which frequently extends beyond the sale of personal skills and services outside working hours to the provision of services during working hours), the use and sale of public equipment, the acceptance of bribes to provide key services or equipment, the sale of legally or illegally imported commodities and other legal and illegal financial transactions, to the sale of legally produced handicrafts, produce from urban allotments, renting of accommodation, and so on (Kenedi, 1981, pp. 44–6; Marrese, 1981).

The position of the agricultural worker in the state or co-operative sector is similar to that of the industrial worker; he receives an income in the form of wages or as a share of the co-operative farm income according to the amount of work performed, which will be supplemented by sales of produce from private plots in the legal free markets. In addition, certain sectors of the population may depend entirely, or mainly, on private-sector incomes (eg, licensed private-sector artisans in Hungary and private farmers in Poland).

Under these circumstances, the quantity and quality of effort provided by the individual, to a specific task and in total, will be affected by such 'Benthamite' considerations as the unpleasantness of the work performed and the compensation to be received.

Private-sector earnings may impinge on the supply of effort to the public sector by affecting the amount of overtime worked, the willingness to undertake employment that leaves little time or opportunity for market activities, and may also affect the value of income differentials and monetary incentives within the public sector, as the worker may be expected to compare the additional income to be obtained through the acquisition of industrial skills, performing unpleasant work, and so on, with the possibility of acquiring similar income in the private sector.

Conversely, the certainty and stability of income generated from employment in the public sector and the range of welfare benefits may also affect the supply of labour to the private sector, particularly in agriculture. For example, the agricultural worker may feel that the marginal benefit to be derived from livestock breeding on private plots will no longer compensate for the considerable additional effort when a satisfactory minimum income can be guaranteed from state and co-operative farm work.

Causes of Disequilibrium in the State Retail Sector

Equilibrium in the state sector will occur when the supply and demand for all commodities is equated. If the state determines the quantities of commodities to be produced, neoclassical analysis predicts that the simplest way to achieve equilibrium is to adjust retail prices (by raising or lowering turnover taxes or subsidies) so that all individual markets are cleared. An additional advantage of this system would be that the level of turnover tax or subsidy could function in a similar fashion to profits in a market economy, by providing planners with an indication of the demand (or lack of demand) for individual commodities, while changes in prices would similarly act as a guide to planners to raise or lower output.

In practice, however, the state is frequently unwilling to allow all markets to be cleared in this way for a combination of social, political and economic reasons. In certain instances the state itself may exhibit a set of social preferences which it attempts to impose on the market as a result of which the consumer may find his individual preferences frustrated, leading to the emergence of a formal or informal rationing system and the possibility of individual preferences spilling over into secondary markets. Many of the cases are not unique to centrally-planned economies and may be observed (although probably to a lesser degree) in mixed economies. The most common examples of the imposition of state preferences involve the pursuit of distributional goals that either cannot be, or have not been realised through wage policy, including the attempt to allocate certain goods and services such as housing, education, hospital and medical facilities according to need, rather than according to work performed (or ability to pay), involving either the free distribution of such facilities or their provision at greatly subsidised prices.

The principle is extended in most East European countries to provide a greater coverage of facilities provided below cost including accommodation, creches, children's clothes, basic foodstuffs, public transport, which will help to redistribute income towards lower-income earners in general (provided such commodities constitute a greater proportion of the expenditure of lower-income families) and towards larger families in particular, thus achieving a more equal per capita income distribution than could be realised by wage policies alone.

In addition, the state may attempt to encourage the consumption of certain marketed commodities which it considers to be beneficial to consumers (milk, nutritious food, and so on) by keeping prices deliberately low, or may try to discourage consumption of commodities that are considered harmful (for example, tobacco, alcohol) by imposing high levels of turnover taxation. In this case the optimum policy for the state is to supply the quantities that are deemed desirable and to choose the prices that will equate supply and demand.

In other cases the pursuit of distributional goals will result in conflict between state and individual preferences. The state will either have to provide sufficient quantities to clear the market at selected prices, or a formal rationing system will be required to ensure that those in need actually receive the goods and services concerned.

In practice the problem may be very complicated. Many goods and services simultaneously satisfy basic needs while providing additional satisfaction — housing provides comfort, convenience and privacy in addition to shelter; education satisfies aesthetic as well as purely social or vocational needs; clothing provides decoration and an indication of status as well as warmth; lower-quality foodstuffs can be used as the ingredients for spiced and flavoured dishes or as feedstocks to animals. Furthermore, the state's policy objectives in the area of distribution and subsidisation are frequently not clearly defined.

It becomes highly unlikely therefore that the system of centralised allocation, in combination with a monetary system, will lead to a direct coincidence between the individual consumer's monetary demands and the supplies made available at the prices determined by state preferences, and as a result a decentralised network of redistribution of goods and services will arise unless suppressed by central authorities. Consumers may attempt to supplement state education facilities by hiring teachers privately, or where permitted, to engage in private house construction which in turn will generate a large number of subsidiary market activities; state housing may be sublet, bribery may be required to obtain state facilities, and so on. Commodities sold at below market-clearing prices may be bought up and resold on black markets, sales assistants may be able to demand bribes before selling commodities, while additional quantities of commodities may be provided legally or illegally in free markets.

A further set of disequilibria in the state retail sector results from errors in the process of planning and producing consumer goods. Output targets based on the previous period's production levels may mean that initial output targets given to enterprises will not reflect changes in consumer demand, particularly those arising from market saturation, while the lack of incentive to innovate at enterprise level applies with particular force to the production of new commodities that have not been specified in output plans. Furthermore, the absence of freedom of entry into large-scale production will prevent those who may have identified the possibility of changes in consumer demand from exploiting their ideas — a constraint that is extended to foreign producers by the foreign trade monopoly.

In the attempt to achieve bonuses linked to plan targets, enterprises frequently have little incentive to produce consumer goods in the quantities, qualities, styles, sizes and colours that consumers demand, but will produce an assortment of commodities that most easily satisfies the output plan. The process of storming may also result in commodities of low quality being produced by a tired workforce towards the end of the planning period in order to fulfill output plans, while other attempts to compensate for the delivery of inadequate components may also lead to low-quality outputs (Nove, 1980, pp. 90–102; Kornai, 1959, pp. 117–46).

The replacement of gross output targets by success indicators linked to profitability and sales will have no impact on these problems if enterprises are still credited with revenue for the production of output regardless of whether a final sale is made in the retail market or not.

East European governments also display a marked reluctance to increase retail prices even in the face of substantial changes in supply and demand conditions. Gomulka (1982a) proposes that this is system-induced. In market economies the cause of price increases is more obviously diffused throughout the economic system and the government is less likely to be held responsible for specific price increases. Montias (1981) also argues that cheap or subsidised prices are frequently seen by the population as a tacit agreement between the Party and the working class. Retail price increases therefore represent a unilateral breach of that agreement which may stimulate wider unrest.

Unfortunately, long periods of price stability may require sudden discrete changes in retail prices to restore equilibrium and the government may be prevented from announcing or discussing these increases in advance, for fear of hoarding. Sudden sharp increases without consultation may stimulate the very unrest it is desired to avoid, while simultaneously providing a focal point for other grievances to be raised.

In Poland attempts to increase food prices in 1970 and 1976 were accompanied by strikes and riots and in 1980 led to the emergence of Solidarity. The announcement of food price increases in the week before Christmas in 1970 indicated considerable government insensitivity, and the rescinding of price increases by Gierek meant that food prices had considerable political significance. The reaction to the July 1980 announcement that certain cuts of meat were to be transferred from the lower-price state stores to higher-priced commercial stores — which implied an increase of 90 per cent to 100 per cent in the price of 2 per cent of all meat sales and a reduction of about half a per cent of real income (Portes, 1981) — also appears to indicate that the lack of consultation and the breakdown of trust in the government and support for its policies were at least as important as the price increases themselves. Similarly, labour unrest in the wake of food price increases of 20 per cent and energy price increases of 43 per cent in Hungary in

1979 appear to have resulted from attempts to dismiss workers in the steel industry as much as from the price increases themselves.

Finally, the retail sector itself displays considerable shortcomings. In part these result from the low priority attached to the act of selling which is regarded as both unproductive and demeaning in Marxian tradition and in consequence fails to attract skilled personnel, while according to Nove (1980, pp. 266–7) the sector suffers from 'low priority, low pay, low morale, underequipment, underfinancing and shortages of suitable premises'. The second major feature is the chronic sellers' market for many commodities which so affects the balance between customer and salesman as to result in rudeness and indifference on behalf of staff in stores, restaurants, and so on.

Polish commentators attribute poor service to shortages of products, shops and staff and the lack of a service tradition. The provision of 'thank you money' to obtain service is frequently mentioned in accounts of the Hungarian retail network (Marrese, 1981). Kenedi (1981, pp. 89–93) argues that 'when sales assistants show great cordiality they expect a bribe'. The problem of bribery in Hungary has received serious attention in academic and trade journals and has been estimated to amount to over ten billion forints a year (or over 3 per cent of total personal income). Kenedi (1981, p. 99) argues that the process of corruption is so prevalent in the Hungarian retail network that it has effectively led to the elimination of queues which have been replaced by an elaborate network of backdoor contracts — ' "We haven't got it" really means "it's under the counter" . . . Getting "No" for an answer simply means: in this particular shop, the assistant either couldn't or didn't want to play the game.'

The existence of lengthy queues in other parts of Eastern Europe has been a fairly common phenomenon, and time spent queueing considerably increases the number of hours of what is effectively unproductive work that must be undertaken to obtain fairly commonplace articles.

Turcan (1977) carried out an interesting investigation of the causes of queues in Poland, on the basis of his observations during three visits to Wroclaw in the period 1974–6. The last visit was undertaken before the price increases of 1976 were announced, and in all periods one would anticipate a degree of excess demand for food products resulting from specific Polish conditions. The supply conditions in Wroclaw were probably worse than those in major cities but better than those prevailing in the country in general. Not surprisingly, the longest queues were for meat, with queues of over 100 people observed on occasion. Queues for food products in general were longer than those for non-food products, although these were also commonplace. With the exception of meat, there were no obvious signs of shortages and stocks of food products were still available when the stores closed.

Other problems that Turcan observed were a fairly general lack of desire on behalf of sales personnel to sell commodities to customers, the sudden closure of apparently adequately stocked stores and a general lack of availability of aids to quick service such as pre-packed foods, baskets in supermarkets, which together with badly laid-out shelves appeared to have the purpose of keeping customers away from commodities. He concluded therefore that with the exception of meat, most queues resulted from organisational problems rather than price disequilibria. Furthermore, he felt that contrary to popular opinion there were sufficient staff on hand to serve customers, but that few were concerned with selling. This he attributed to the fact that sales personnel were financially responsible as a group for any losses resulting from dishonesty, and were therefore financially motivated to prevent theft and to ensure that customers were always supervised and had little direct access to commodities.

Secondary Markets

Differences between state and individual preferences, errors in planning and production and deficiencies in retailing will all result in a lack of coincidence between consumer demands at the prices they are willing to pay, and the quantity and prices of goods and services provided in the state sector. This therefore provides an opportunity for subsidiary market activity to equate supply and demand.

The position again may bear many resemblances to the role of market activities in mixed industrial economies with large state and corporate sectors. A perfectly legal form of market activity here frequently involves smaller operators filling the lacunae left by large organisations that are unable, or find it uneconomic in terms of their organisational structure, to perform a range of small services. These may be most apparent in such areas as retailing, servicing consumer durables, small-scale construction, and household repairs. In many cases, by offering external economies to one another the two sectors may exhibit a high degree of interdependence, although in the longer run the corporate sector may try to incorporate smaller operators by, for instance, direct subcontracting.

In capitalist economies illegal market activity may be stimulated by the desire to avoid government regulations, taxation, and so on, which allows a smaller operator to keep his private financial costs below those of rivals who are obeying legal requirements, while 'classic' black market behaviour frequently involves placing purchasers in contact with supplies of commodities for which there is excess demand at prevailing prices, as a result of administrative rationing or distribution (eg, tickets for sporting events).

All of these activities may be expected to occur at the points of interaction between the market and state sectors in a centrally-planned economy, resulting in legal and illegal market activity. There is however one critical difference between the interaction of the sectors in the two systems. In most cases in the capitalist economy, the active nature of money means that the secondary market economy can obtain its supplies of inputs (including labour) by perfectly legal methods which may even sustain the state and corporate sector.

In the interaction between the state and market sectors of a centrally-planned economy, the inactivity of money prevents the market sector obtaining the supplies it requires from the state sector, and there will be no *a priori* reason to expect a better coincidence between planned supply and demand than occurs in the state retail sector.

A contrast may be drawn here between legal private agriculture in Poland and the private sector in Hungary. Much of the apparent failure of private agriculture in Poland can be attributed to its inability to obtain basic inputs such as specialised implements, small tractors, and appropriate fertilisers, while these facilities have been directed towards state farms. Preference towards the state sector has adversely affected the performance of the private sector, although this sector itself is vital for the efficient functioning of the Polish economy in total. A similar relationship of interdependence prevails between the two sectors in Hungary where it appears that the private sector is more successful in drawing off resources from the public sector, to the extent that this may even have detracted from the performance of the public sector. These effects may be most noticeable on the quality of labour supply and the use of state materials for private gain.

Gabor and Galasi (1981) have estimated that the equivalent of one million man-years of labour per year are spent in the legally approved sectors of private agriculture and private construction in Hungary compared with a total economically active population of 5.2 million people. Many of the hours worked in agriculture are however performed by pensioners, students and others who would not have been seeking active employment in the state sector and therefore must be considered as supplementary to work provided in the state sector. Furthermore, many workers are estimated to work about three hours a day on private plots in addition to their state sector occupation.

It has been calculated that average earnings in the private sector are about 100 forints an hour compared with 19 forints in the public sector (Volgyes, 1981). A major incentive to work in the public sector appears to be that it is frequently a prerequisite for obtaining profitable work in the private sector with consequent deleterious effects on the quality of labour supplied in the state sector.

Private agricultural plots are normally only available to the families of those with primary state-sector employment. (As a result 1.8 million small plots provide subsidiary employment to 5 million people, including pensioners on a part-time basis totalling up to about 0.8 million man-years per year) (Gabor and Galasi, 1981). The fiscal regulations for those registered as full-time private-sector artisans (plus the difficulty of obtaining registration) mean that it is frequently more lucrative to perform private work as an unofficial supplement to state-sector income.

The most important reason for taking state-sector employment as an adjunct to private-sector employment is the necessity (and financial advantage) to be obtained from utilising state-sector materials in the secondary economy. The advantages of this practice have been most clearly described in the construction industry by Kenedi (1981), who describes a range of illegal (but unofficially approved) activities taking the form of borrowed capital equipment for private-sector use (including large-scale equipment such as lorries and drills), as well as stolen inputs such as bricks, sand, cement and tiles. Furthermore, state-sector employment may be the best method for finding potential customers and much labour time in the state sector is reported to be spent seeking outside employment opportunities to the detriment of the quality of 'official' work. Moonlighting often takes place during formal working hours in addition to work in the evenings, at weekends and during holidays, which may leave the worker too tired to perform his official work adequately. Gabor and Galasi also argue that workers take sick-leave to pursue private-sector employment and that the continuity of labour to the state sector is interrupted by periods of employment in the private sector which can amount to 10 per cent of registered work time.

Although the incidence of private work is likely to be greater in the legally approved house construction sector in Hungary — amounting to 120,000 man-years according to Gabor and Galasi (1981, p. 48) — Marrese (1981, p. 50) cites Hungarian authorities to demonstrate that such activities extend not only to occupations such as architects, plumbers and painters, but also to doctors, lawyers, teachers, cooks and mechanics.

Such secondary economy activities will affect both the efficiency of the state sector and interpretations of the efficiency of the economy based on official statistics. If, for example, the proportion of unrecorded secondary economy activity increases relative to recorded activity, official growth rates will underestimate the real rate of growth, while increases in the use of state capital and raw materials for unrecorded purposes will be reflected in an apparent decrease in the efficiency of the economy as a whole, measured in terms of unit of recorded input per unit of recorded output.

Many of the problems of secondary economy activity and their repercussions on the 'official economy' apply also to market economies. The possibility of maintaining a higher income through tax evasion, legal and non-contractual moonlighting, including the use of employers' facilities and equipment are by no means unknown in market economies, and will affect the use of material incentives and the use of piece-rates on the supply of labour to specific industries, and to the 'official economy' as a whole.

Some qualitative distinctions between the two systems may be made however. Firstly, the absence of active money for the supply of inputs to the legal private sector makes such illegal activity a necessity for the successful operation of the private sector. More crucially however, a far greater role for government policy in centrally-planned economies is exercised through its control over the state sector. A secondary economy, particularly one that offers wage rates that are five times higher than the official economy and in which income tax is negligible, will mean that attempts to influence income distribution through wage policy will be considerably affected by market sector activities. As far as the consumer is concerned the secondary economy does provide a range of personal services that are frequently unobtainable in the official economy, while it is frequently reported that secondary-sector earnings are essential to maintain a satisfactory standard of living. This last fact tends to imply that turnover tax levels in the state sector are set sufficiently high to cream off some elements of private-sector earnings.

Hard Currency and Coupon Stores

The subject of multi-tier price systems and the existence of separated markets cannot be left without some mention of the existence of foreign-currency stores and coupon stores for senior Party personnel. Foreign-currency stores provide an outlet for the sale of imported or higher-quality domestically produced consumer items that are only available for foreign currencies. These stores are nominally available for use by foreign visitors, or by citizens who have legally acquired foreign currency (eg, through remittances from overseas, personal income from work performed or sold overseas, a proportion of which may be retained or exchanged for certificates for use in foreign-currency stores or through central allocations). The status value attached to such commodities frequently gives rise to black markets for foreign currencies and certificates, where the exchange rate may be several times higher than either the official rate of exchange or the tourist or commercial rates.

In addition, 'coupon stores' exist in the USSR and East European countries, in which senior Party officials and other members of the elite can obtain higher-quality commodities in exchange for certificates that are not available to the general public. Moreover, members of the elite frequently benefit from other non-pecuniary allocations such as accommodation, specialised hospital facilities and educational privileges for their children. Money (in the form of currency) is effectively 'inactive' in this sector.

The elite only receive these privileges for the period they remain in office — the system therefore provides a pervasive method of control.

6 MACROECONOMIC EQUILIBRIUM AND THE CONTROL OF INFLATION IN EASTERN EUROPE: THEORY

Introduction

It is apparent from Chapter 5 that microeconomic disequilibria can and frequently do arise in the state markets for goods and services in Eastern Europe as a result of differences in the supply and demand for specific commodities at centrally-determined prices. Excess supply in individual markets is reflected by the existence of stockpiles of unsaleable items at prevailing prices, while excess demand is reflected by the existence of queues outside state stores for commodities in short supply, official and unofficial rationing and the emergence of legal and illegal markets.

The evidence of supply shortage raises the question of whether excess demand exists in the aggregate, at prevailing price levels in the state markets and whether supply shortages should be regarded as a symptom of inflationary pressures.

The importance of the distinction is that if shortages result predominantly from excess demand pressures in the aggregate, they would be eliminated by general price increases, or by reductions in consumers' disposable incomes. Although in the short run each individual consumer might perceive himself to be worse off, as a result of a reduction in his nominal income, his real purchasing power will remain unchanged, while many of the very real frustrations and time losses resulting from queueing, engaging in black market activity, hoarding commodities, and so on, would be eliminated. In addition the avoidance of such time losses, and a greater coincidence between monetary incentives and the availability of commodities to be purchased with extra income, could in itself result in greater productivity and hence increased real income.

If, on the other hand, consumer shortages result largely from the specific microeconomic disequilibria outlined in Chapter 5, reductions in consumers' nominal income will also mean a reduction in real income and will be largely ineffective in reducing specific shortages and queues, which will require precise measures including changes in relative prices, improving the responsiveness of enterprises to consumer demand, improving retailing, and so on. Reductions of real incomes under these circumstances are likely to result mainly in further hardship and unrest.

A further possibility is that the two phenomena may coexist — a generalised excess demand may be accompanied by the institutional problems outlined above, resulting in the need for separate policies to tackle both problems.

Before examining the mechanics of maintaining consumer equilibrium in the aggregate, some critical institutional factors which help to simplify the task of central authorities in maintaining macroeconomic equilibrium require examination.

The Absence of Real and Money Multipliers in the State Sector[1]

The passivity of money in the state sector means that changes in money demand will not result directly in changes of output, which will in turn lead to further changes in income, which will generate further changes in demand. While money is passive, a purchase by a consumer in a state retail store has a similar macroeconomic effect to the payment of a sales tax or excise duty in a market economy. This does not apply in the case of purchases in (legal or illegal) free markets which create income for the recipients and in turn create additional demand.

The significance of idle money balances in the Stalinist economy is the opposite of that attributed to them by the Keynesian analysis of deficit demand, but is in many ways analagous to the system of compulsory savings subsequently espoused by Keynes (1940) in 'How to Pay for the War'. Marx described the hoarding of money which would prevent the generation of capital investment as 'sheer tomfoolery' (Walker, 1978, p. 180–3). This analysis is reflected in Soviet practice by the passivity of money in the investment sector. Investment is determined by central planners responding to real not money variables and seeking to maintain a level of aggregate demand consonant with full employment. The principal macroeconomic role of the State Bank is to finance that investment, and the Keynesian explanation of deficit demand caused by financial institutions holding idle money balances in order to speculate, or insure against falls in the value of fixed assets, should not arise.

Money balances in the traditional Soviet system are partly made up of small sums to maintain enterprise liquidity, but mainly represent the private savings of households held in the form of deposits at savings banks or in the form of cash balances which act as a store of value. The principal problem in analysing the economic significance of these savings is to estimate whether they represent voluntarily or involuntarily deferred demand.

The Separation of the Domestic Price Structure From Foreign Trade Prices (the Preisausgleich)[2]

The domestic price structure is established independently of the price of imported inputs. Prices of imported goods are converted into units of domestic currency at an appropriate rate of exchange, and an adjustment is made from the state budget (in the form of a tax or subsidy) to equate import prices with domestic prices. A similar process takes place in the case of exported commodities. A major advantage of this process from the macroeconomic viewpoint is that changes in the prices of imported goods do not feed directly into the domestic price level, and the economy can be effectively isolated from imported inflation.

A distinction must be made between the effects of isolation from world inflation (ie, generalised changes in world price levels) and isolation from changes in relative world market prices. If the prices of all imports and exports are rising at a uniform rate, budgetary equilibrium will be reattained by corresponding adjustments to the rate of exchange at which foreign prices are converted into domestic prices. Similarly, if relative prices are unchanged there will be no need to adjust domestic production decisions, while in the case of small or temporary changes in relative prices it may be felt that the benefits of domestic price stability outweigh any efficiency losses. This will not be the case if relative world prices are substantially affected. If investment is still appraised in terms of domestic prices, planners will have no incentive to develop advantageous export or import-saving projects, or to discontinue investments that require relatively expensive imported components or raw materials. Similarly, even if enterprises are affected by price criteria the separation of domestic from foreign trade prices will make them unresponsive to changing world conditions.

Furthermore, if export prices rise more slowly than import prices, worsening terms of trade will mean either that planners will have to run a balance of payments deficit or cut back on domestic production as a result of reduced resource availability. A deflationary domestic policy requires cutting personal consumption and/or central expenditure (involving cuts in defence, health, education, and so on). Unless the impact is borne entirely by central expenditure, domestic equilibrium will require either cuts in wages, increases in personal taxes, price rises or increases in voluntary savings. Under most of these circumstances the anti-inflationary benefits of the separation of domestic from foreign prices may prove to be illusory.

Consequently, although the *Preisausgleich* may have offered some benefits to the East European economies in helping them to isolate the effects of generalised inflation in world markets in the period up to the early 1970s, the costs of continuing this policy after the energy crisis may well have outweighed the benefits. Hungary is currently reaffirming its intention to proceed without the *Preisausgleich* and more recently Romania has announced the intention to feed world market prices directly into enterprise wholesale prices.

The attractions of the *Preisausgleich* for imported consumer goods may remain. The policy of utilising the foreign trade monopoly to raise budget revenue by importing consumer goods and selling them at appreciably higher prices on the domestic market was suggested by Preobrazhensky in 1921 (Day, 1975), and is reported to have been advocated by Stalin personally to Romanian planners seeking advice on how to combat excess demand pressures in the 1950s. The deflationary effects of this policy are identical to those of a tariff and mean that excess demand can be siphoned off without affecting the price level of basic commodities. In addition, a substantial overpayment of wages in the domestic sector can be equilibriated by a smaller balance of payments deficit, while the fact that consumers can only obtain certain commodities in this way provides an incentive to accumulate large money balances, possibly in excess of annual income, and to respond to material incentives.

The Mechanics of Maintaining Macroeconomic Equilibrium in the Consumer Market

The separation of monetary circuits into a cashless circuit for inter-enterprise transactions and a monetary circuit for the payment of wages means that the problem of macroeconomic equilibrium is reduced to equating consumers' disposable incomes minus their savings, to the value of available consumer goods and marketed services. The working of the foreign trade monopoly and the

Preisausgleich also mean that the analysis of macroeconomic equilibrium in the consumer goods market can be simplified by including imported consumer items as items of domestic consumption and evaluating them at their domestic sale price.

In the absence of free markets, the task of maintaining consumer equilibrium in the aggregate in the state sector can be explained by the following equation:

(1) $$Y_1 + TR - S_1 - T_X = P_1C_1$$

Where Y_1 = earned income in the state sector.
TR = transfer payments received by individuals from the state.
S_1 = savings from state-sector income.
T_X = taxes and other payments to the state (excluding turnover tax).
P_1 = price level of consumer goods (including turnover tax) and services marketed by the state.
C_1 = quantity of consumer goods and services marketed by the state.

(1A) and where $$Y_1 = W_M L_M + W_{NM} L_{NM}$$

Where W = wage rates in the state sector.
L = quantity of labour employed in the state sector.
Subscript M = in the production of marketed output.
Subscript NM = in the production of non-marketed output.[3]

Alternatively, as wages in the state sector are paid to workers producing both marketed and non-marketed commodities (investment, defence, education, health, free goods and services, overseas aid) macroeconomic equilibrium also requires that the volume of personal savings, personal taxes and turnover taxes are equated to the wages paid for the production of non-marketed commodities plus transfer payments.

(1B) $$TR + W_{NM} L_{NM} = S_1 + T_X + TT_1$$

(Exports may be considered to be non-marketed output and the receipts from the sale of imported consumer goods as turnover tax.)

Excess demand can result from a number of factors including

overpayment of wages in the state sector (Y_1), resulting either from overpayment of wages (W_M and/or W_{NM}); overfulfilment (ie, above plan employment) of labour in the state sector (L_M and/or L_{NM}), increased transfer earnings (TR), pensions, grants, child and maternity benefits, and so on; failure to fulfil output in the marketed sector (C_1); disproportionate growth of the wage fund relative to output in the marketed sector; or a shift of output from the marketed to the non-marketed sector.

Equilibrium would be restored by an equivalent increase in savings (S_1), personal taxes (T_X) or turnover taxes (TT_1). Increases in turnover taxes will lead to an increase in retail prices (P_1) which, if accurately recorded in the retail price index, would show up as open inflation.

If insufficient changes are made in personal taxes or turnover taxes, the gap between consumers' incomes and expenditure will have to be filled by increased savings. If the latter are higher than those consumers actually planned to make at current prices and incomes, they should be described as involuntary savings and be taken as an indication of repressed inflation.

The introduction of free markets (both legal and illegal) complicates the process, as consumer expenditure in free markets also creates incomes for the recipients. Consequently, increases in prices in free markets will be insufficient to choke off excess demand generated in the state sector, as they will in turn create additional income, which will reappear as demand. A process analagous to the Keynesian multiplier in the retail sector could operate, with excess state-sector incomes generating increases in free market incomes which can either be saved or spent in state stores or in free markets.

Although expenditure from free market incomes in state stores would result in that amount of money income being withdrawn from circulation, it would also lead to an equivalent reduction in the amount of commodities on offer — consequently excess demand in the state sector would not be choked off. Recipients of state-sector income would still find attempted purchases in state stores frustrated and would be forced either to save their income in anticipation of goods becoming available, and/or to spend that income in free markets. It is, however, likely that recipients of free market income will choose to save some of that income, while personal taxes will have to be paid on legal private-sector income. Equilibrium will only be restored when the initial volume of excess demand created in the state sector is matched by an equivalent increase in savings and personal taxes out of all incomes.

Equilibrium will now require:

(2) $$Y_1 + Y_2 + T_R - S_1 - S_2 - T_X = P_1C_1 + P_2C_2$$

Where Y_2 = income from free market activities.
S_2 = savings from free market incomes.
P_2 = price level of free market activities.
C_2 = quantity of free market goods and services.

There is no *a priori* reason to suppose that the velocity of circulation will be the same in free markets as in the state sector, and a situation could arise whereby a small amount of excess demand in the state sector generates a number of rounds of expenditure in free markets, which would in turn require substantial rises in free market prices before equilibrium is restored. As the authorities frequently apply legal maxima to free market prices which are below the level which would restore equilibrium (and certain market activities are illegal for economic rather than social reasons), the possibility of excess demand at prevailing legal prices still exists.

The Causes of Excess Demand in the Consumer Sector[4]

Clearly it is possible for temporary disequilibria to arise between aggregate supply and demand in consumer markets because of exogenous factors beyond planners' control (eg, a bad harvest). It is necessary to ask whether planners correct for such imbalances in subsequent periods, and whether there are any systemic reasons why excess demand should be generated in a centrally-planned economy. Wiles (1968, pp. 47–52) argues that excess demand pressures result from the planners' refusal to bankrupt a losing enterprise and are, therefore, systemic. If the State Bank is faced with the prospect of an enterprise that cannot fulfil its output targets without violating its cost plans, Wiles proposes that it will give priority to the output plan and sanction above plan payments.

If the labour plan is the constraint to the fulfilment of the enterprise output plan, this will result in cash being advanced directly for the payment of above plan wages, and provided that labour plans are not underfulfilled elsewhere in the economy, this will immediately generate excess consumer demand. This situation will arise when planned productivity levels have been set too high and enterprises succeed in employing more labour than allocated in the plan, by drawing additional workers into the industrial labour force, a phenomenon that was particularly noticeable during the first Soviet five-year plan. Alternatively, if priority is given to the non-marketed sector (especially investment in heavy industry and defence) labour may be diverted away from the production of consumer goods. Furthermore, if the enterprise cannot attract sufficient quantities of labour to fulfil the plan at planned wage rates, it may attempt to attract labour by paying higher rates by such devices as the deliberate falsification of work norms, artificial upgrading, and overstatement of overtime.

If the material input plan is the constraint to output plan fulfilment the State Bank may make a cashless advance to the enterprise in order to allow it to obtain above plan inputs. In order to provide the additional inputs the supplying enterprise may overfulfil its output plan, and make perfectly legitimate above plan payments to labour in the form of bonuses, piece rates, and so on. Consequently, a non-cash injection in one enterprise may generate a cash credit elsewhere, and if the cash credit does not result in a cash withdrawal in the first enterprise, excess demand will arise with no additional commodities to satisfy that demand. Finally, any additional payments for overfulfilment of targets in the non-marketed sector will result in short-term excess consumer demand.

Any overpayment of the wage fund relative to the availability of state-sector commodities will result in inflationary pressures, unless it is matched by an equivalent increase in savings or taxes out of private and state-sector income combined.

Types of Inflationary Pressure

These inflationary pressures can be divided into three potentially coexisting forms:

(a) *Open inflation* — Inflation revealed by changes in the official retail price index of commodities sold in state stores. This should, in theory, be revealed by an increase in P_1 in equation (2). In practice, however, the official retail price index cannot include all items sold in state stores leading to the possibility of:
(b) *Hidden inflation* — price increases that actually take place but are not reflected in the official price index, either due to deficiencies in the construction of the index itself (ie, it does not accurately measure prices actually paid in state stores) or due to changes in prices in secondary markets (P_2 in equation 2) that are not

reflected in the official index, or due to changes in the proportions of commodities that consumers buy in the two markets.

(c) *Repressed inflation* — price increases that would take place if markets were allowed to clear themselves, resulting in a *generalised* shortage of commodities at existing prices in relation to purchasing power and reflected by a higher volume of savings ($S_1 + S_2$ in equation 2) than that planned by consumers at existing prices and incomes. Nuti (1981a) shows that Polish economists distinguish between two types of excess demand that contribute to this category — an 'inflationary gap' which is the difference between intended and actual purchases out of current income in a given time period, and 'inflationary overhang', which consists of the excess monetary assets in the hands of the population and is composed of the sum of all the preceding periods' inflationary gaps. Consequently, the elimination of inflationary gaps alone will not restore full equilibrium to the economy if 'inflationary overhang' exists. Equilibrium will only be restored by reducing the purchasing power of existing savings either by a confiscatory monetary reform, or by increasing prices more than incomes, or by such a substantial increase in both prices and incomes that existing money balances are devalued.

Hidden Inflation

There is a considerable volume of anecdotal evidence which indicates the existence of hidden inflationary pressures in the Soviet and East European economies, which has caused many western analysts to doubt whether official price indices accurately measure the rate of change of prices that consumers *actually* pay in all markets to acquire consumer goods and services. In all countries price indices are approximations and suffer from considerable methodological and conceptual problems concerning the weighting to be attached to individual items, the prices at which they should be evaluated, and the base period over which they should be measured. Consequently, many of the more general criticisms applied to Soviet and East European price indices apply also to western statistics.

There are, however, some significant methodological criticisms which are specific to the calculation of price indices in the USSR and several East European countries — the most inclusive being that they are not based on sample data which reflect the quantities purchased at the prices actually paid, but are based on a fixed basket of standard commodities, regardless of availability, evaluated at centrally-determined prices which are contained in official lists (Bornstein, 1972; Schroeder, 1975). Consequently, as state retail stores cannot *officially* alter prices to reflect changing demand conditions, the level of open inflation is effectively what central authorities *say* it is. Unless those central authorities vary the level of turnover tax to equate supply and demand and record the resulting changes in official indices, the level of open inflation will not be an accurate measure of total inflationary pressures in the economy, and comparisons between rates of open inflation amongst East European economies may purely reflect differences in the flexibility of central planners' responses to changing demand conditions, and differences in the accuracy of their statistics.

One set of criticisms argues that official price indices do not accurately reflect the prices and quantities of commodities sold in state and co-operative stores. These criticisms include:

(a) The appearance of higher-priced new or improved commodities that differ little, if at all, from the lower-priced commodities previously on offer which are then removed from sale (Bush, 1973; Schroeder, 1975; Katsenelinboigen, 1975).

(b) Commodities are sold in state stores at prices higher than those on official lists (Birman, 1978; Schroeder, 1975).

(c) Persistent shortages of commodities in state stores, including the non-availability of cheaper items, which force consumers to buy higher-priced substitutes, or to buy the identical or similar commodities in higher-priced free markets (Katsenelinboigen, 1975; Schroeder, 1975).

(d) Bad service in retail stores, accompanied by sales assistants demanding bribes or favours before serving customers, or charging for a service not performed (Schroeder, 1975).

(e) The exclusion of commodities for which prices have increased from official retail price indices (Birman, 1978; Bornstein, 1972).

A second set of criticisms refers to the existence of a multi-tiered price system which the official indices make no attempt to capture. These criticisms include:

(f) The existence of legal free but controlled markets largely for agricultural commodities and handicrafts in which prices are higher than those in state stores (Schroeder, 1975).

(g) The existence of illegal black markets for the resale of goods and services at prices higher than those existing in state markets (Simes, 1975).

Portes (1977) argues that the phenomena listed above are all symptoms of disequilibrium in the consumer market which may force consumers to accept less-preferred substitutes for the commodity that they originally intended to purchase, or even to buy a commodity at a price higher than they had originally intended to pay. All of these micro disequilibria could however coexist with equilibrium in the consumer market in aggregate. Consequently, he argues that if the consumer substitutes the same quantity of preferred lower price commodities by alternatives at a higher, but unchanging price in each period (cases c, f and g), or pays the same bribe to receive service in each period (case d), these are not in themselves proof of hidden inflation. For hidden inflation to occur, either the prices of commodities not included in the official retail price index must rise from one period to another, or the proportion of total consumption obtained by means of higher-priced substitutes must increase from one period to another. Finally, the existence of a multi-tier price system is not necessarily indicative of hidden inflation, but could result from deliberate social policy. Portes (1977, p. 121) cites Hungarian statistical evidence to show that 'hidden' price increases for new goods amounted to only 1 per cent per annum in the 1950s. Portes (1977, p. 121) also shows that free market food prices in Hungary and Poland increased at the same rate as state retail prices for food between 1960 and 1968, a trend that continued in Hungary after the introduction of the New Economic Mechanism, but his table also shows that in Poland free market prices rose by 44 per cent from 1968 to 1975, while state retail prices increased by only 5.4 per cent. It is noticeable that during this period total state wage payments grew substantially less quickly than state retail turnover in Hungary (largely as a result of open inflation), whereas in Poland state wage payments outgrew the money value of retail turnover from 1971 to 1975 as the state imposed price controls. Inevitably therefore, unless the ratio of voluntary savings from state income increased, some of this purchasing power would leak to the private sector.

Table 6.1: The Growth of Household Savings Deposits in State Banks, 1976–80

	Rudcenko's data	Deposits of households in state savings banks							
		Annual growth rate					As % of state retail turnover		
	1975[a]	1976	1977	1978	1979	1980	1970	1975	1980
Bulgaria		7.9	4.9	3.5	7.9	7.3	70.9	92.7	86.1
Czechoslovakia	72.5	8.9	8.6	3.8	3.2	5.4	39.2	54.9	61.1
GDR	104.2	6.5	4.9	6.9	5.4	2.8	80.6	90.8	98.1
Hungary	62.1	14.2	15.7	16.2	8.7	6.9	30.0	37.0	42.2
Poland	57.8	10.3	10.9	10.3	11.6	7.9	25.5	36.5	35.8
Romania		16.0	16.9	24.5	11.9	14.2	na	32.5	46.7
USSR		13.1	13.1	12.4	11.5	7.0	30.0	43.3	57.9

Note: a. Total money holdings of the population as a percentage of state retail turnover.

Sources: Column 1: estimated from Rudcenko (1979); Columns 2 to 8: estimated from CMEA statistics (1981); also, for Romania, IMF (October 1981).

Repressed Inflation and the Growth of Savings

Observers of the Soviet and East European economies frequently draw attention to the rapid growth of private household savings held in official state savings banks and in the form of idle money balances over the last twenty years. Although the rate of growth of total money wages and incomes has been approximately matched by the rate of growth of state retail turnover, household deposits in state savings banks have grown 2½ to 5 times faster, and over the period from 1960 to 1975 have grown at rates ranging from 330 per cent in the GDR to 1770 per cent in Poland (Table 7.1).

There is a considerable body of opinion amongst western analysts and Soviet emigré analysts that these savings are involuntary and represent repressed inflation (Birman, 1980a, 1980b; Katsenelinboigen, 1975). Their argument is that consumers are prevented from making their desired purchases out of current money incomes at existing prices in state retail stores by the non-availability of commodities in aggregate, while central controls prevent state retail prices from rising to meet money demand. Although this excess money demand may filter through to private markets (and be a source of *hidden* inflation), price controls in free markets, and legal restrictions on the supply of many commodities (and in particular the purchase of inputs to manufacture private commodities) prevent free markets from fully equilibriating supply and demand. Consumers are therefore forced to accumulate idle money balances in excess of their desired holdings at prevailing prices and incomes because of the non-availability of commodities and central restrictions which prevent prices from equilibriating supply and demand.

This view has been questioned in a series of econometric studies conducted by Portes and Winter (1977, 1978, 1980) who argue that the disequilibria observed in the consumer markets in the Soviet Union and Eastern Europe resulted from the microeconomic problems described in the preceding chapter. Excess demand for some commodities at prevailing prices coexisted with excess supply of other commodities due to the failure of central authorities to set market clearing prices for *specific* commodities, not from excess demand in the aggregate. A corollary of this argument is that the observed levels of savings represent voluntarily deferred demand.

It is important to get the growth rate of savings into some perspective. Firstly, the more spectacular growth rates result from very low initial volumes of savings, largely as a result of confiscatory currency reforms enforced in the post-war period. In Hungary and Poland, for example, household deposits in state savings banks were 4.0 per cent and 4.8 per cent respectively of gross money incomes at the end of 1959.[5] Moreover, in every country for which data exist the growth rate of savings deposits in banks has been faster than the growth rate of total household money balances. This is consistent with a pattern of households with low initial income levels holding idle balances purely to finance current consumpton and subsequently putting savings into interest-earning bank deposits (to finance the purchase of durables) as income grows. Furthermore, as more members of the population move beyond basic subsistence incomes the growth rate of savings in aggregate will exceed the growth of the average individual's savings.

Finally, the quantity of savings is a *stock*, while income is a *flow*, and comparisons which relate the growth of income levels to the growth of the stock of savings may give a misleading impression (see Portes, 1977). (A household could, for example, save 5 per cent per annum out of a *static* income and achieve a substantial growth of its stock of savings.) Comparisons which relate *changes* in income levels to *changes in the annual flow* of new savings indicate that savings levels in Eastern Europe may not be as excessive as they initially appear. It can be demonstrated purely arithmetically (taking figures that are fairly representative of East European experience) that a household saving a constant 7 per cent of an income that grows at 7 per cent per annum would accumulate a stock of savings equivalent to 80 per cent of its annual expenditure at the end of fifteen years.

Using more sophisticated techniques of calculation than the example given above, Portes and Winter (1977, pp. 13–14) have estimated marginal propensities to save of between 6 and 8 per cent out of absolute income for Czechoslovakia, the GDR, Hungary and Poland. They argue, on the basis of a comparison with international rates of saving out of disposable income, that instead of showing signs of forced or involuntary savings 'the long run savings ratios of the CPE's are relatively low'.

In a later study Portes and Winter (1980) question the existence of repressed inflation in CPE's by attempting to estimate whether consumer markets are characterised in the aggregate by excess demand or excess supply. In a CPE the combination of centrally-determined prices and quantities of consumer goods, passive money, and the state monopoly of foreign trade, will mean that any divergences from money demand generated in the state sector will not provoke an automatic adjustment in supply in the form of changes in domestic or imported supplies of consumer goods, or their prices. Similarly unplanned changes in supply will not lead to automatic changes in demand (by depressing incomes) or to changes in the price of consumer goods, or in the volume of imports. Consequently, until central planners make the appropriate adjustments, the actual level of consumer purchases in any period will be determined by the *lower* of desired consumption expenditure or the available supply of consumer goods. The existence of chronic repressed inflation would therefore be reflected by a persistent excess of money demand at prevailing income levels over the supply of consumer goods at prevailing prices, leading to accumulating volumes of money savings.

Portes and Winter (1980) tested the probability that actual aggregate consumption levels in all markets were determined predominantly by the level of supply (ie, excess demand), or the level of demand for those East European economies for which adequate data on household income expenditure and savings could be obtained (Poland and Czechoslovakia from 1955 to 1975, the GDR from 1957 to 1973 and Hungary from 1957 to 1975). Annual demand for consumer goods was estimated as being inversely related to the level of household savings in the preceding year, and positively related to the level of disposable income in the preceding year and changes in the level of disposable income in the year in question. They proposed that the planned supply of consumer goods fluctuated around a long-term growth trend according to changes in households' liquid assets and short-term fluctuations in output. In the case of Poland and the GDR an estimate of the effect of changes in expenditure on investment and defence 'crowding out' the supply of consumer goods was included.

Excess demand was apparent in only 33 of the 78 annual observations and existed as the dominant explanation of consumer disequilibrium in the GDR only. Poland was found to have deficient money demand relative to supply in the 1960s, followed by substantial excess demand in the early Gierek years of 1971 to 1973 which was eliminated by price rises and *imported consumer goods* in 1974 and 1975. In Czechoslovakia deficient demand in the early 1960s peaked in 1967 and was followed by excess demand until 1973. In

Hungary the dominant regime of deficient demand from 1964 to 1975 was broken only in 1968 and 1970. These results differed somewhat from their initial results (Portes and Winter, 1977) and the significance of some of the years in question is referred to in the next chapter.

Portes and Winter (1980, p. 156) concluded that statistical problems prevented them from drawing 'strong, unambiguous empirical conclusions' but they believed that 'the evidence clearly justifies rejecting the hypothesis of sustained repressed inflation in the market for consumption goods and services' for the four economies over the time period under consideration.

They conclude that planners do make some form of adjustment to changes in demand and supply and have successfully attempted to combat excess demand in the consumer sector, and that the disequilibria observed there are caused principally by micro-economic disturbances. Although there was some evidence that the demands of the investment and defence sectors crowded out consumption in Poland and the GDR, the consumption sector is not used purely as a residual sector when other supply and demand factors are altered. Portes and Winter also consider (1980, p. 156), but reject the possibility, that their estimates of demand (based on the previous period's levels of savings and income) would fail to capture *continuous* excess demand, even as low as 3–5 per cent of consumption expenditure throughout the period.

Unfortunately, their studies embrace the period from 1955 to 1975 only, and cannot reflect the period after 1975; they also only measure domestic consumer equilibrium rather than aggregate demand and supply pressures that might spill over to the foreign trade sector. Consequently, the possibility that excess demand pressures in the economy are satisfied by sucking in a quantity of imports that cannot be sustained in the long run is also consistent with their findings. Similarly, domestic equilibrium in the aggregate may be difficult to sustain following a deterioration in the terms of trade. If this should result in reduced domestic resource availability, while planners are unable to cut state expenditure or money wages, some combination of open, hidden and repressed inflation will occur. This hypothesis is examined in the next chapter which may help to explain the attempts to eliminate repressed inflation subsequently experienced in Poland and Romania.

Portes' and Winter's arguments do appear to be at variance with those of Nuti (1981a) who argues that excess demand has existed in Poland throughout the last decade, and that the fact that money balances have grown faster than the sales of goods and services, combined with evidence of a flight out of domestic currency, are evidence of a growing gap between desired and actual money balances, indicating repressed inflation (which he defines as an increase in the degree of excess demand). Consequently, even if imports of consumer goods in 1973 and 1974 succeeded in restoring a balance between aggregate demand and supply as Portes and Winter suggest, they need not have eliminated the 'inflationary overhang' generated from 1970 to 1973. If however they purely *reduced the level* of excess demand generated in that year the amount of inflationary overhang would have still increased. The resumption of a trend where incomes grew as fast as sales would therefore result in an increased level of excess demand.

Nuti (1981a, p. 6) argues that these pressures are general throughout Eastern Europe and that 'shortages of basic consumption goods persist and their effect is becoming cumulative, successive inflationary gaps raising the inflationary overhang'.

The Incentive to Save

Although the East European marginal propensity to save may appear to be low in comparison with that of many Western industrial nations, there are reasons to suppose that the incentive for the individual to save may be lower in a socialist economy than in a capitalist economy.

Conventional analysis of western savings behaviour assumes that the principal motivations for the individual to save (ie, to defer consumption to a later period) are to achieve a greater volume of consumption at a later period (by receiving interest on money lent to investors, who use it to acquire resources to increase the productive capacity of the economy) and to compensate for fluctuations in income and desired expenditure that may be expected throughout one's lifetime. The individual, then, may purchase private insurance to cover actuarial risks or save to cover them out of his own accumulated wealth, resulting in saving to support a family, to finance education or anticipated health expenditure or to finance reductions in earnings resulting from old age, incapacity through ill health, or the vicissitudes of market demand for one's labour, and so on. Even if such savings are lent to private financial intermediaries (for example, banks, building societies, insurance companies), they will normally be accounted as personal income not spent on current consumption.

In the aggregate such savings by one sector of the community should be offset by dissaving elsewhere, but when the economy is growing it is probable that personal savings out of current income exceed dissaving. (Young and middle-aged people may anticipate a higher life-time income and save more for their retirement than older generations who are currently dissaving; they may demand better education facilities for their children than they themselves received, and improved health facilities for their old age than their parents could afford, and so on.)

In the socialist economy the state bears a much greater responsibility for financing productive investment and social expenditure through its taxation policy, *before* this reaches the individual in the form of personal income.

The reduced need for personal savings in the aggregate is channelled back to the individual by the greater financial security provided by the state, which results in a lower necessity to set aside current income for precautionary reasons and by lower opportunities to invest for capital gain.

The former results from the provision of welfare services such as free or low-cost health and education facilities, low-cost housing, old age and disability pensions, maternity and family allowances, and provision of nursery facilities and creches. In addition the greater degree of job security, the relative equality of money incomes (by western standards) combined with reduced opportunities for temporary, high windfall earnings reduce both the source and need for money savings (Ofer and Pickersgill, 1980). The practice whereby the benefits available to those in elite occupations consist largely of the provision of facilities and services, and the opportunity to buy scarce commodities exists only while they retain office, means they have less opportunity than their western counterparts to accumulate high money savings for expenditure after their period of relative prosperity has ended.

The lower opportunity to establish private businesses, to acquire venture and risk capital and the lower returns on private investment, even for house purchase, also result in a reduced incentive to save out of personal income. These factors also reduce the need and incentive for savings by the rural population (who traditionally save a higher proportion of income than the urban population in market economies due to the greater potential variability of income and the need to reinvest) on state farms who receive guaranteed wages and do not finance investment.

It could be anticipated that savings ratios would be higher in those countries that permit a greater degree of market activity (eg, Hungary) where income fluctuations and the incentive to save to establish businesses, buy housing, and so on, would be greater. Furthermore, if, as Ofer and Pickersgill (1980, p. 125) propose, savings out of private income are larger than those from state-sector income, the multiplier effect in private markets may not be large.

Under these circumstances, it is difficult to predict *a priori* what a 'normal' savings ratio out of current personal incomes should be in East European CPE's. Ofer and Pickersgill (for the USSR) and Portes and Winter (for four East European CPE's) propose that savings behaviour in CPE's can be explained by the same hypotheses and variables as are considered appropriate for western market economies, and that Soviet and East European savings ratios are accordingly lower than those to be found in market economies.

This argument is rejected by Birman, mainly on the grounds outlined above (1980b). He places considerable emphasis on savings for investment purposes in market economies, whereas the major cause of saving in a CPE is deferred consumption. With very limited credit facilities, individuals put aside current income to acquire consumer durables and to meet retirement needs that cannot be covered by pensions. The critical indicator of the importance of savings, he argues, is not its rate of change through time but its absolute level in relation to what is available in retail markets. Birman (1980a, p. 89) estimates that by the end of 1976 private money balances in the state savings banks and private hoards in the USSR were the equivalent of 66 per cent of the money income of the population, 84 per cent of state retail turnover and over three times the stock of available commodities to meet that demand.

Comparable estimates of total money holdings as a proportion of expenditure in the state and co-operative retail network in 1975 based on the data used in Portes' and Winter's studies are shown in Table 6.1. They demonstrate that the absolute volume of money holdings represented more than six months' purchases in each country, but were a lower proportion of state retail turnover than Birman's estimate for the USSR for all countries except the GDR. Figures showing the growth of household savings deposits in state savings banks and expressing these as percentages of state and co-operative retail turnover are also shown in Table 6.1. Rudcenko's data (1979) show that these are a considerable underestimate of total household balances and that the proportion of total savings held in state savings banks varies from country to country and from time to time within the same country. The data do however show a progressive growth in the stock of savings held in banks in relation to retail turnover in each country (except Poland) in the years preceding major retail price increases (Bulgaria, 1980; Hungary, 1979; Romania, 1982). In Poland and Romania the annual growth of bank deposits has exceeded 10 per cent in each year since 1975. In Romania, the growth of wage rates of 10.6 per cent in 1978 was accompanied by savings bank growth of 24.5 per cent. In Hungary and Bulgaria the growth of the stock of savings was slowed down by price increases, and in the GDR and Bulgaria deposits in savings banks alone were the equivalent of nearly a year's turnover.

These represent substantial amounts of deferred demand, which as Birman (1980b) indicates are almost entirely held in liquid form. Even assuming such holdings were acquired voluntarily (which Birman dismisses) the total level of liquid money balances still provides a critical problem for central planners attempting to cut back on current consumption in view of deteriorating terms of trade and the need to repay foreign loans. Birman (1980b, p. 589) also suggests 'if (when) the population demands its money back a political catastrophe may occur'. The comment was addressed to the Soviet situation but is not an inappropriate description of events in Poland.

Even if the repercussions are not so drastic elsewhere, the implications of idle balances substantially in excess of those the population wishes to hold are severe both for the operation of East European economies and for the preservation of social stability. If workers find they cannot convert wages into consumer goods at the rates they desire, monetary incentives, wage differentials, and so on, become ineffective. In the short term it becomes more worth while to seek to spend existing balances than to acquire extra money income. Time is more profitably spent for the individual by queueing for consumer items, in following up rumours of goods availability, in secondary market activity and hoarding what is available, further decreasing supplies and productivity and fuelling unrest. In the longer term, monetary incentives to improve labour productivity, to acquire new skills, to attract labour to the new green field sites, where social and cultural facilities are basically lacking, will be virtually ineffective. Reform measures placing greater emphasis on decentralisation and market stimuli will also be of little effect.

The problems are very similar to those that stimulated the debates in War Communism, the NEP, and the first three years of the first five-year plan in the USSR. On the one hand there may be greater pressure for the use of non-monetary incentives (for example, the provision of social and cultural facilities attached to the workplace, better-stocked factory retail stores in key areas) to attract labour to specific places — a process that may see the expansion of a form of Socialist Corporatism. On the other hand, the restoration of money incentives would require monetary reform that would be overtly or tacitly confiscatory. The latter could be achieved by successive price and wage increases which destroy the real value of money balances, and this appears to be the course currently being pursued in Romania and Poland. In the interim this could increase popular unrest as the population tries to divest itself of devaluing assets and is not without considerable political dangers.

Notes

1. See Wiles (1968), Chapter 3.
2. See Wiles (1968), Chapter 6.
3. The analysis of the distinction between the wages paid in the marketed and non-marketed sector is due to Gregory and Stuart (1974).
4. See Wiles (1968), Chapter 6 and (1979), Chapter 14.
5. Estimated from Rudcenko (1979). Rudcenko's data were the statistical basis for the study by Portes and Winter.

7 WAGE PRESSURES AND OPEN INFLATION IN EASTERN EUROPE

Introduction

The period of rapid industrialisation in the USSR in the 1930s was also marked by high levels of open inflation. Recent Soviet estimates indicate that retail prices in state and co-operative stores in 1932 were 250 per cent of the 1928 level and had doubled again by 1937 (Garvy, 1977, pp. 172–7). There is a broad consensus that inflationary pressures were not the result of deliberate policy, but arose from defects in the operation of the system of credit control and money supply which allowed both wage rates and the volume of industrial employment to rise significantly faster than planned, while resources were simultaneously concentrated in heavy industry. The major achievement of macroeconomic policy in this period was that passive money prevented the overpayment of the wage fund from diverting resources from investment to consumption. Although industrialisation was de facto financed by inflation, there is little evidence that this was the result of deliberate intention rather than financial slackness. Between 1948 and 1953 the USSR embarked on a policy of reductions in state retail prices while increases in salaries and wage rates were prohibited (Katsenelinboigen, 1975). This is consistent in Marxian theory with improvements in productivity which lead to reductions in socially necessary labour being reflected in a reduced value of the commodity concerned. It is also directly analagous to the theoretical working of a market economy under conditions of perfect competition where productivity improvements should reduce the average total cost of production and hence the price of output (as new entrants to the market compete away the supranormal profits of existing producers). The policy of annual price reductions was ended after Stalin's death on the grounds that this did not give any incentive to producers to improve their productivity.

Soviet official indices still reflect a remarkable degree of price stability between 1955 and 1977 with the official state retail price level in 1977 approximately equal to its 1955 level, although evidence of open inflation can be seen after that date.

There are some parallels between Soviet and East European experience. Considerable open inflationary pressures were experienced in Eastern Europe in the immediate post-war period, the most spectacular case being Hungary, where the price level increased by 3.81 to 10^{27} times between August 1945 and June 1946 (an average monthly increase of 19,800 per cent). This was accompanied by a policy of direct expropriation whereby 1,000-pengo notes were only acceptable when accompanied by stamps to the value of 3,000 pengos, thus reducing private wealth holdings (Wilczynski, pp. 233–4).

Although interpretation of data on the quantities and value of retail trade turnover is subject to a number of difficulties, there is evidence of open inflationary pressure in all East European countries from 1948 to 1955. Studies of the Polish system of monetary control by Farrell (1975) and Podolski (1973) attribute this to the inability of the central bank to administer stringent controls of the wage fund (while resources were being concentrated towards investment in heavy industry and towards meeting demands generated by the Korean War) rather than to deliberate inflationary policy. Inflationary pressures in the 1948–55 period therefore bear more resemblance to Soviet experience in the 1930s than to that of War Communism.

There are some further parallels between the behaviour of East European official price levels and those of the USSR. Official price indices for Bulgaria, Czechoslovakia and Romania all show an average open inflation rate of under 1 per cent per annum from the late 1950s until the early 1970s, while the GDR shows a negative inflation rate (Table 7.1). Poland alone of the East European countries has shown an open inflation in each year since 1956, averaging more than 2 per cent per annum up to 1972.

Table 7.1 demonstrates the patterns of growth of aggregate demand and supply for commodities in the state retail networks and illustrates some of the underlying features of domestic macroeconomic control in the period 1960–75. Rows 3 and 4 show that the growth of gross money wages in the state and co-operative sector was fairly closely related to the growth of the value of retail sales in the state and co-operative network throughout the bloc. (In Romania alone the growth of wage payments exceeded the money growth of retail sales by 1 per cent per annum.) Rudcenko (1979) has provided detailed estimates of the growth of disposable money income (composed of all monetary incomes from the socialised

Table 7.1: The Growth of Income, Wages, Expenditure and Savings, 1960–75 (1960 = 100)

	Bulgaria	Czechoslovakia	GDR	Hungary	Poland	Romania	USSR
A. Aggregated indices							
Gross money incomes	320	241	175	314	369	na	na
Disposable income		237	175	308	391	na	238
Goss wage payments	387	223		248	400	412	298
(state and co-operative sector)							
State retail trade							
(1) Total turnover	340	225	183	322	368	351	268
(2) Real turnover	307	208	189	264	296	331	268
(3) Price level	111	108	97	121	122	106	100
Money holdings							
(a) All sources	na	545	387	838	1030	na	na
(b) Savings banks	950	601	430	1478	1870	na	829
B. Individual statistics							
Average wage	186	169	na	182	243	212	181
Number of employees	207	132	133	137	165	194	165
Per capita retail turnover	306	207	187	303	320	304	225
Real per capita turnover	276	192	193	250	262	287	225
Purchasing power of average wage in state stores	167	156	na	150	199	200	181
Index of real wages	166	149	na	153	169	176	161
Real income per head (All sources)	232	na	184	199	239	227	197

Notes: Gross money incomes: all incomes from the state sector plus private-sector earnings (except Poland). Disposable income: gross money income minus taxes, social security contributions, etc. Gross wage payments: average wage in state and co-operative sector multiplied by number of employees. State retail trade: official retail trade in state and co-operative stores. Total Turnover in Current Prices: real turnover obtained by deflating by official price indices for retail turnover. Money Holdings (all sources): includes cash held by households, savings banks, holdings of households in state savings banks.

Sources: Gross money income (except Bulgaria), disposable income (except USSR), money holdings (all sources) calculated from Rudcenko (1979). Disposable income, USSR, from Portes (1977). Gross money income, Bulgaria, extrapolated from Portes (1977). Gross wage expenditure, state retail trade, savings banks deposits and Section B all calculated from CMEA statistics (various years).

sector plus, with the exception of Poland, private-sector earnings, minus taxes, social security contributions and other obligatory payments to the state) for Czechoslovakia, Hungary, Poland and the GDR, which also show a close correspondence with the growth of state retail turnover.

Rows 5 and 6, calculated from official price indices for trade in state and co-operative retail stores, show that the major part of the growth of retail trade was officially described as a real increase, while price increases for all countries except Poland and Hungary were substantially less than 1 per cent per annum.

In each country a significant factor in the growth of money wages has been increased employment in the state and co-operative sector (row 10). This corresponds to the pattern of extensive growth and has been primarily achieved by increased participation rates in the state sector, rather than through increases in the population of working age. The growth of employment in the state sector has been larger in those countries with a higher pre-war proportion of the labour force occupied in agriculture and with higher estimated disguised unemployment in agriculture. In Bulgaria, for example, over half of the growth of money wages from 1960 to 1975 resulted from increases in the labour force.

The effects of this are reflected in the rates of growth of consumption in the individual economies. The growth of real consumption in the state sector as a whole (row 5) and per head of the population (row 12) has been highest in the more agrarian countries of Bulgaria and Romania, where the growth of the state labour force has been highest and has been lowest in the industrial economies of the GDR and Czechoslovakia where there has been less potential for transfer out of agriculture. Similarly the countries with an intermediate stage of development, Poland and Hungary, have had intermediate rates of growth of the labour force and per capita consumption in the state sector.

A consequence of the pattern of extensive growth, involving increased participation rates in the economy and the transfer of labour from the less productive to the more productive sectors of the economy, is that individual wage rates have grown more slowly than average family income and consumption. The extent to which the individual household benefits from economic growth will depend therefore on the ability of household members to transfer to higher-paid occupations and/or increase their supply of labour in total. A greater proportion of the population will benefit materially from extensive growth in localities where increased employment opportunities are available to family members, and where the possibility of transfer to better-paid jobs is higher. The principal beneficiaries will be young mobile workers, particularly those moving from agricultural to industrial employment (who may remit part of their incomes to their homes) and families whose personal circumstances (age, location, possibility of mothers and young adults entering the work force) permit them to increase their labour supply.

The growth of real incomes in families where the number of wage earners is constant and who possess few possibilities of transferring to higher-paid occupations will be considerably below the national average, and their satisfaction with the rate of growth of wages and with the economic system and the government may be expected to be influenced accordingly. Occupational mobility will be lowest in areas that are heavily dependent on established industry (mining and extraction, steel, shipbuilding, and so on) and where the problem may be compounded by a low growth of work opportunities for family members.

The implications of this analysis are serious for East European leaders who derive much of their support from traditional working-class areas and occupations and who will find that dissatisfaction with material progress will first be experienced in just those regions.

A situation can then arise which resembles the analysis of inflationary pressures in the United Kingdom provided by Jackson, Wilkinson and Turner (1972). They propose that occupations can be graded on a continuum from high to low (or negative) *growth* of productivity. Such differences in productivity *growth* may depend on factors beyond the control of the individual worker, such as the provision of improved machinery, economies of scale (which will result in decreased labour productivity as industries contract) and improved organisation. Under conditions of perfect competition such productivity changes in the unit cost of production would be passed on to the community as a whole in the form of price changes for final products. In economies where imperfect competition exists in the product and labour markets, there is a strong resistance to price and wage reductions (the Keynesian ratchet) while workers seek to benefit from economic growth by securing pay increases linked to *individual* productivity (a process that is frequently legitimised by income policies). Consequently, workers in low productivity *growth* sectors do not benefit to the same degree from economic growth as a whole and tend to press for commensurate wage increases to preserve their relative income position. Such claims are frequently legitimised by arbitration tribunals who place emphasis on equity, preservation of differentials, and so on. Wages and salaries in the aggregate, therefore, grow faster than productivity, resulting in inflationary pressures.

In the initial round inflation appears to be the vehicle of distributive justice. If, after a period of time workers in the higher productivity growth sectors come to expect inflationary pressures and attempt to secure real income increases equivalent to productivity increases, they will press for money wage increases higher than the growth of productivity (or economic growth in the aggregate) generating an income-cost spiral.

How such a process evolves will depend on institutional factors including the willingness of employers to grant wage increases and on the fiscal and monetary policies of the governments concerned. Under East European circumstances the following institutional circumstances may be noted: continuous price reductions as a means for passing on productivity increases have been rejected since the mid-1950s; employers, trade unions, monetary and fiscal authorities are all directly responsible to the government; the government determines (or exerts direct pressure) on wages and prices and should therefore be empowered to prevent such pressures arising.

It was argued above that those in a position least likely to benefit from income growth or increased productivity in Eastern Europe are those on whom leaders have traditionally depended for support, and who are best suited by the nature of their work and by their geographic concentration to press successfully for money wage increases in excess of productivity, which may spill over into unrest if not granted.

Granting such wage increases will not result in inflationary pressures if income growth in other sectors of the economy is contained. If central planners fail to do this, however, the resulting increases in demand will lead either to open inflation (caused by increases in turnover tax to absorb demand) or (if prices are controlled) to suppressed inflation.

There is some evidence to suppose that these pressures have occurred in East European economies in the 1970s and that central authorities may be initially reluctant to take the measures necessary to bring about equilibrium. In certain cases governments have been forced to cede to labour unrest focused in areas with strong working-class traditions and to pressures to resist real income changes (Poland, 1970, 1976 and 1980; Romania, 1977 and 1981). In decentralised economies where bonuses have been linked to enterprise profitability (Czechoslovakia, 1967–8 and Hungary since 1968) complaints have been made about the unfairness of income distribution, particularly to workers, which have been compensated by wage increases and open inflation, leading to pressures for recentralisation.

The East European countries have also experienced a substantial deterioration in their terms of trade in the 1970s resulting from increased energy prices, which has reduced the resources available to the domestic economy to satisfy demand increases.

Finally, as more labour has been absorbed into the more productive sectors of the economy and participation rates approach ceilings determined by demographic factors, the potential for increasing incomes from labour mobility has been diminished and an increasing proportion of the population has come to depend on the rate of growth of individual wage rates as the major source of growth of family income. Consequently, the rate of growth of real personal household income can only be maintained from domestic sources if the rate of growth of labour productivity increases or if resources are shifted from the non-marketed to the marketed sector. Planners may be reluctant to adopt the latter course as the major components of non-marketed output are investment, which will affect future growth, and non-monetary components of family real income (such as schools, creches and hospitals), which will directly affect family welfare. Consequently, if authorities yield to the temptation to allow family income to grow at rates maintained in earlier years, without a sufficient growth in productivity to offset deteriorating terms of trade they will be forced to run balance of payments deficits and/or incur domestic inflationary pressures (open or suppressed).

Statistical Evidence

The relationship between money incomes derived from employment in the state and co-operative sector and expenditure in the state and co-operative retail network is examined in Table 7.2. The estimated level of open inflation in state stores is given by column

(f) for each country, while a slightly different official price index is given for Hungary in column (g). In addition, for countries where reasonable data on total money income and/or real personal money incomes are available, these have been estimated in columns (h) and (j). While column (f) confirms the low level of open inflation experienced by each country of the bloc during the late 1950s and 1960s, it also indicates that open inflation was being experienced in each country of the bloc by the late 1970s and that Hungary and Poland have been undergoing open inflation since the early 1970s.

Open inflation has been manifested to a different degree and at different times in the countries of the bloc due to differences in economic factors such as the level of development and the potential for increasing labour force participation, the dependence on, and source of, imported energy supplies and the cost at which they are obtained, and because of differences in political factors or economic policy choices, including the willingness of the leadership to acquiesce to wage and income pressures, the willingness and ability to cover domestic demand pressures from imports, the degree of decentralisation of the economy, and so on.

Table 7.2 shows that until the early 1970s the rate of growth of the real volume of turnover in the state retail sector — column (e) — exceeded the growth of money wage rates — column (a) — in each country, due to the growth of labour supply to the state sector — column (b) — which resulted in total wage payments in the state sector — column (c) — growing substantially faster than wage rates. Macroeconomic equilibrium in the state sector therefore required that the money value of retail turnover — column (d) — should grow faster than wage rates to equate total wage payments, a practice that was achieved largely by increases in output rather than by increases in prices as a result of increased labour supply to the consumer goods sector. Personal real money income — column (j) — therefore rose faster than wage rates. This pattern has not been sustained in the 1970s as the rate of growth of the labour force has declined and terms of trade have deteriorated. In each East European country, at some stage in the 1970s, the real volume of retail turnover has grown more slowly than average wage rates (and therefore total wage payments) leading to open inflationary pressures, and culminating in declining growth (and even on occasions a decline in the absolute level) of real personal incomes. Average wages have grown faster than the real volume of state retail trade in

Table 7.2: Earnings and Expenditure in the State and Co-operative Sectors, 1960–80

Annual average rate of growth of:

(a) Wage and salary rates (average monthly)
(b) Employment in state and co-operative sector (excluding collective farms)
(c) Total wage payments in state and co-operative sector — (b) × (a)
(d) Retail turnover in state and co-operative sector — current prices
(e) Real volume of state retail turnover
(f) Prices in state retail stores
(g) Official price index — if different from (f)
(h) Net money incomes of the population
(i) Real wages
(j) Personal real money income

Bulgaria

	(a)	(b)	(c)	(d)	(e)	(f)
1961–5	3.4	4.4	7.9	8.5	7.2	1.2
1966–70	6.0	4.6	10.8	9.4	8.6	0.7
1971	2.2	4.2	6.5	6.5	6.6	−0.1
1972	3.5	4.5	8.2	6.5	6.5	0.0
1973	6.3	9.3	16.2	9.0	8.8	0.2
1974	2.0	4.6	6.7	9.7	9.2	0.5
1975	3.1	7.4	10.7	8.2	7.9	0.3
1976	1.2	5.7	7.0	7.6	7.3	0.3
1977	2.2	−0.4	1.8	3.5	3.1	0.4
1978	3.9	0.6	4.5	4.7	3.7	1.0
1979	5.1	1.3	6.4	7.2	2.2	4.9
1980	17.0	−0.5	16.5	18.0	3.6	13.9

Romania

	(a)	(b)	(c)	(d)	(e)	(f)	(g)
1961–5	5.5	5.8	11.6	10.3	10.5	−0.2	
1966–70	5.2	3.5	8.9	7.4	7.2	0.2	
1971	2.6	5.2	7.9	9.5	8.8	0.6	
1972	1.8	4.7	6.6	6.1	6.1	0.0	
1973	4.3	3.6	8.1	7.9	7.1	0.7	
1974	6.4	3.3	9.9	10.5	9.3	1.1	
1975	9.2	4.6	14.2	8.6	8.4	0.2	0.2
1976	5.7	4.1	10.0	9.5	8.7	0.7	0.5
1977	7.2	2.8	10.2	6.9	6.4	0.5	0.5
1978	10.6	3.2	14.1	11.4	10.3	1.0	2.0
1979	4.8	3.3	8.3	7.1	6.0	1.0	1.8
1980	6.2	2.2	8.5	7.6	5.6	1.9	1.5

Hungary

	(a)	(b)	(c)	(d)	(e)	(f)	(g)	(h)[a]	(i)	(j)
1961–5	2.3	2.4	4.8	5.3	5.2	0.1	0.5	2.3	1.8	3.1
1966–70	4.2	2.4	6.7	9.6	8.8	0.6	1.0	4.5	3.5	6.4
1971	4.7	1.0	5.7	9.2	7.4	1.7	2.2	4.6	2.3	4.2
1972	4.4	2.5	7.0	6.5	3.3	3.1	2.8	5.1	2.2	3.0
1973	6.9	1.5	8.5	9.4	5.8	3.5	3.4	6.3	2.8	4.5
1974	6.7	1.5	8.3	12.0	9.2	2.1	1.7	7.4	5.6	6.4
1975	6.2	1.3	7.6	10.0	5.4	4.4	3.8	7.7	3.8	4.2
1976	5.7	0.5	6.2	6.7	1.4	5.3	5.0	5.1	0.1	0.3
1977	8.9	4.5	13.8	10.5	6.2	4.0	3.9	7.8	3.1	2.5
1978	8.1	0.7	8.9	8.9	4.0	4.7	4.6	7.8	3.1	2.5
1979	6.3	−0.2	6.1	11.6	1.7	9.7	8.9	7.0	−1.7	−0.3
1980	6.2	−1.2	4.9	8.4	0.5	7.8	9.2	7.2	−1.8	−0.9

a. *Net* nominal wages.

Czechoslovakia

	(a)	(b)	(c)	(d)	(e)	(f)	(h)	(i)	(j)
1961–5	1.8	2.2	4.0	3.4	2.7	0.7	4.8	1.1	4.1
1966–70	5.3	1.9	7.3	7.9	6.1	1.7	7.6	3.5	5.8
1971	3.7	0.8	4.5	5.2	5.5	−0.4	5.5	4.1	5.9
1972	4.1	1.3	5.5	4.9	5.3	−0.4	6.0	4.5	6.4
1973	3.3	1.3	4.6	6.0	5.8	0.2	6.4	3.1	6.2
1974	3.3	1.2	4.5	7.5	7.1	0.4	4.6	2.9	4.2
1975	3.2	1.1	4.3	3.4	2.8	0.6	3.8	2.6	3.2
1976	2.8	1.3	4.1	3.8	2.6	1.2	4.6	1.6	3.4
1977	3.2	1.1	4.3	3.9	2.6	1.3	4.6	1.9	3.3
1978	3.0	1.2	4.2	5.5	3.5	1.9	3.5	1.0	1.6
1979	2.5	1.1	3.6	3.6	0.0	3.6	3.6	−0.9	0.0
1980	2.4	0.8	3.2	2.0	−0.7	2.7	n.a	−0.3	n.a

Poland

	(a)	(b)	(c)	(d)	(e)	(f)	(h)	(i)	(j)
1961–5	3.7	3.5	7.3	7.1	4.7	2.2	7.6	1.4	5.2
1966–70	3.7	3.3	7.1	7.1	5.4	1.6	7.6	2.2	5.2
1971	5.5	3.1	8.8	8.4	9.7	−1.2	11.0	5.7	12.3
1972	6.4	4.4	11.1	11.9	11.9	0.0	15.3	6.4	15.3
1973	11.5	3.9	15.8	12.8	9.9	2.6	14.5	8.7	11.6
1974	13.8	3.5	17.8	14.6	7.3	6.8	14.9	6.6	7.6
1975	11.8	2.3	14.4	14.1	10.8	3.0	13.2	8.5	9.9
1976	8.8	1.0	9.9	13.8	8.7	4.7	11.5	3.9	6.5
1977	7.3	1.4	8.8	13.0	7.7	4.9	12.4	2.3	7.1
1978	6.1	0.6	6.7	8.6	0.0	8.7	8.7	−2.7	0.0
1979	8.8	0.1	8.9	10.0	3.1	6.7	9.9	2.4	3.0
1980	13.5	0.0	13.6	9.5	0.9	8.5	9.8	3.9	1.2

Sources: All figures have been estimated from data initially provided in the official statistical publications, economic journals and government statements of the countries concerned. Attempts have been made to reconcile the numerous discrepancies between the various sources to arrive at consistent figures, particularly where index numbers have been used as primary sources. I am confident that the figures represent a reasonable guide to the broad patterns indicated by CMEA statistical sources but would be reluctant to use them for precise cross-section econometric measurement.

National statistics have been supplemented by the following sources: CMEA statistics (various years) especially for columns (a), (b), (c), (d), (e), (f); Rudcenko (1979) column (h) up to 1975; Nuti (1981a) for Poland, columns (a), (e), (f), (h) after 1970; Gomulka (1982a), Poland, column (i); Altman (1980) Czechoslovakia, column (h) 1975 to 1979; IMF (October, 1981) Romania, column (g); Hungary — taken entirely from national statistics.

Hungary in each year since 1972 (except 1974), in Poland since 1973 (except 1977), in Czechoslovakia since 1975, in Bulgaria since 1978 and in Romania since 1977 (except 1979).

The initial impetus to inflationary pressures appears to have occurred in the second half of the 1960s in the period following domestic reforms. In every country the proportion of the rate of growth of total wage payments attributable to the rate of growth of wage rates as opposed to the rate of growth of employment, increased in the 1966–70 plan period in comparison with the preceding plan period. This was brought about by an acceleration of wage rate growth in Czechoslovakia, Hungary and Bulgaria (wage rate growth declined only in Romania which did not enact an effective reform) combined with a deceleration of the rate of growth of labour supply in each country except Bulgaria. The increase in wage rates would not be inflationary if accompanied by a commensurate increase in labour productivity but in practice labour productivity growth declined in Bulgaria, Hungary and Romania and remained static in Poland.

At the some time state retail turnover measured in current prices grew faster than in the preceding plan period everywhere except Romania, and Poland where it remained static. Consequently, in 1966–70 most countries experienced a situation in which wage rates and retail turnover in the state sector were increasing, just as the growth of the sources of supply — the growth of labour supply and productivity per worker — were declining.

Country Studies

Czechoslovakia

Czechoslovakia, the most industrially developed country in the bloc, was the first to encounter labour supply problems. In addition to experiencing the slowest rate of growth of labour supply in the bloc in the period 1961–5 Czechoslovakia also incurred the lowest rate of growth of real wages (column i) and real retail turnover (column d). Although the last two factors cannot be attributed to the stagnation of labour supply alone, but must be partially attributed to the lack of specialisation in the CMEA and the fact that Czechoslovakia, as one of the more technologically advanced countries of the bloc benefited less from intra-bloc technology transfer, and was unable to acquire technology from the West, the

Table 7.3: Annual Rate of Growth of Wages, Income and Expenditure in Czechoslovakia, 1966–70

	1966	1967	1968	1969	1970
Wage rates	2.7	5.5	8.2	7.4	3.0
Gross money income	5.6	7.4	11.9	11.4	4.6
Retail turnover					
(a) Money	4.9	6.3	14.0	12.0	2.3
(b) Price level	0.5	2.0	1.3	4.0	1.7
(c) Real	4.4	4.2	12.5	7.7	0.6
Labour productivity	7.4	4.8	5.2	5.3	3.3
National income	10.1	6.9	5.5	6.1	6.0
Imports	2.4	−2.1	14.8	7.1	12.1

Source: *Statisticka Rocenka CSSR* (various years).

slow rate of growth (including a negative growth rate of national income in 1963) gave support to the reform movement.

An initial price reform was announced in January 1966 which mainly affected wholesale prices, but also freed 20 per cent of retail prices from central limits (Kyn, 1975, p. 124). Difficulties involved in estimating the new set of wholesale prices (which increased by 24.2 per cent in 1967) combined with fears that a further decentralisation of retail prices would generate open inflation and stimulate pressure for wage increases (as existing disequilibria were removed) dissuaded the central authorities from their original intention to remove controls on retail prices in (what were intended to be) the first stages of reform in 1967–8 (Kyn, 1975).

Wage indices and limits on the total wage fund were abolished in 1967, and a stabilisation tax levied on the enterprise was introduced, at a rate of 30 per cent on any increment to the wage fund, as the principal method for resisting the pressure for wage growth (Kyn, 1975, p. 127). The measure was largely unsuccessful as can be seen from Table 7.3. Annual figures for this period can be slightly misleading as policies were changed in mid-year, but a fairly clear pattern emerges in which initial increases in productivity were followed by increases in wages and incomes that could not be sustained in real terms leading to inflationary pressures, followed by a subsequent clampdown as centralisation was reintroduced.

The high growth rates of labour productivity and national income achieved in 1966, must in part be attributed to elimination of the inefficiency which had caused the poor performance in the preceding five-year plan period. Consequently, the growth rate of labour productivity and national income dipped in 1967, but wage rates and gross money incomes continued to grow faster than in any earlier year. Real retail trade continued to grow at around 4 per cent and retail prices by 2 per cent. A further round of wage and money income increases took place in 1968 and 1969, and in the uncertainty generated by the Soviet invasion of August 1968 and the installation of the Husak regime the following March many citizens preferred to spend their incomes rather than accumulate cash balances. Retail turnover rose by 14 per cent in 1968, while the import of consumer goods enabled this to be met by a 12.5 per cent increase in real turnover.

Demand pressures led to open inflation in 1969 which was combatted by a partial price freeze on 2 July, which was generalised the following January (Kyn, 1975, pp. 124–8). Simultaneously the authorities pressured enterprises to moderate the growth of wage payments and direct central controls were reintroduced in 1971.

In the next five-year plan the authorities made strenuous and apparently successful attempts to eliminate inflationary pressures. Price controls were extended through 1970 and 1971 and the official index of retail prices registered falls in 1971 and 1972. However, money incomes — Table 7.2, column (h) — grew faster than real retail turnover, resulting in an increase in the savings rate (Rudcenko, 1979). Portes and Winter (1980, p. 151) have estimated that consumer markets in the period 1968–72 were characterised by excess demand. This imbalance was corrected in 1973–5 by reductions in wage rates and personal incomes, and a substantial increase in the balance of payments deficit with the West, much of which was centred on consumer goods.

In the next plan (1976–80) the real growth of retail turnover has not kept pace with the (comparatively low) growth of total wage payments and money incomes, which has led to a gradual escalation of open inflation as the impact of reduced real wages and income growth has been largely transmitted by price increases. By 1979, real retail turnover was static, leading to open inflation of 3.6 per cent, a reduction in real wages and a zero growth of real personal income. Real incomes were again static in 1980.

Poland

Following the Czechoslovak experience in which the low growth of wage rates lent popular support to the reform movement which, in turn, led to the overthrow of Novotny, the reforms of 1967–8 and subsequent Soviet invasion, it may not be surprising that the clearest example of 'expectation induced inflation' initially financed by a balance of payments deficit arose in Poland, the country where real wage growth was the lowest in the bloc in the period 1966–70.

Polish economists have argued that economic development was held back in the late 1960s by the conservative import and growth policies pursued by Gomulka and, in particular, by his reluctance to import intermediate goods and machinery and equipment from the West. Consequently, it was argued that the Polish economy, similar to the Czechoslovak economy in the 1961–5 plan, contained much underutilised potential in this period (Gomulka, 1982b).

The Gierek government was confronted with the problem of gaining popular control under circumstances where the state's authority to determine prices and wages had been successfully challenged. The chosen policy incorporated a 'dash for growth' which involved importing not only machinery and equipment and components, but also consumer goods, which it was intended would act as a stimulus to increased labour productivity by allowing real wages and retail turnover to grow rapidly, while simultaneously holding down prices (Gomulka, 1982b). This policy enabled retail turnover to increase by more than 10 per cent per annum in the period 1971 to 1975 but total wage payments and net disposable incomes — Table 7.2, column (h) — grew substantially faster culminating in inflationary pressures. The desire to avoid consumer price increases in the first years of Gierek's office resulted in the excess demand pressures observed by Portes and Winter (1980). Coincidentally therefore, Czechoslovakia and Poland, with very dissimilar regimes, pursuing different economic strategies, experienced both excess demand and price reductions resulting from very similar initial economic and political pressures.

General increases in state retail prices were gradually reintroduced in Poland in 1973, but unlike Czechoslovakia, the rate of growth of wage rates permitted real wages to grow faster than in 1972, and total wage payments in the state sector rapidly outstripped the volume of both real and money turnover in state stores. Although partly offset by increased personal taxes, net money incomes grew faster than money retail turnover, and money holdings of the population grew by 25 per cent.

In 1974 enterprises used the powers granted to them to increase prices, but these were largely confined to those commodities that were not considered politically sensitive and retail price inflation reached 6.8 per cent, accompanied by a 13.8 per cent increase in money wage rates. Prices of food products were stabilised, but increased subsidies had to be paid to farmers to compensate for increased costs. These subsidies increased from 19 billion zloties in 1971 to 166 billion in 1980 (Nuti, 1981a). A high rate of growth of personal money incomes was maintained in the mid 1970s, and much of the impact of reducing the growth of real incomes was borne by price increases.

As the growth of the labour force declined and the terms of trade deteriorated in the second half of the 1970s, Gierek's policies wavered between attempts to bring about domestic equilibrium at the expense of external equilibrium by continued imports of consumer goods; further unsuccessful attempts to increase food prices (leading to the June 1976 riots) in order to stimulate agricultural production and cut back on consumption; and from 1976 onwards increasing the proportion of national income devoted to consumption. Total wage payments and money incomes have continued to grow faster than real retail turnover, resulting in both open inflation and a growth in the savings rate. In 1978 the real volume of retail turnover remained static and the growth of personal money incomes of 8.7 per cent was entirely swallowed up by price increases, while real wages actually declined. Subsequent increases in money wage rates and incomes exacerbated the domestic imbalance, leading to the attempt to increase food prices in 1980, and the growth of Solidarity.

Industrial wages grew by 27.4 per cent between April 1980 and April 1981, while labour productivity declined by 12.4 per cent (Nuti, 1981a). This process was accompanied by shortages of basic products, lengthening queues and rising prices on legal and illegal free markets. The stock of money balances grew to 1,000 billion zloties at the end of 1981, 33 per cent higher than at the end of 1980. In an attempt to restore financial equilibrium and reduce the real value of money balances, the military government announced food price increases of 241 per cent and domestic heating price increases of 121 per cent in February 1982, accompanied by certain wage increases to help those without money reserves (Nuti, 1982). Although this policy may help towards restoring equilibrium in the retail sector as a whole, it is likely to cause cases of severe individual hardship to those with low levels of savings to fall back on.

Romania

In Romania the slow-down in the rate of growth of the labour force in the state sector in the mid-1960s was reversed as a population bulge of young people born in the early 1950s started to enter the workforce. This factor, combined with imports of machinery and equipment on credit from the West, enabled the policy of extensive growth to be continued, while reform proposals approved in December 1967 were put in abeyance. Unlike Poland, Romania did not allow excess domestic demand to leak through to imports of consumer goods in the early 1970s, and substantial trade surpluses were made in the traditional areas of exports of foodstuffs and basic consumer goods. The provision of western credits did however reduce the necessity to export these items, and it is noticeable that when western governments expressed concern at Romania's indebtedness in 1973, Romania managed a substantial increase in exports of furniture, footwear and clothing, indicating that Romania's apparent inability to sell these commodities in the preceding years may have been partially determined by macro-economic pressures (Smith, 1982b). In the mid-1970s Romania succeeded in containing the growth of its indebtedness, and despite severe floods in 1975 which caused widespread damage to agricultural produce, managed to maintain a rate of growth of retail turnover of 8 per cent.

It was greatly assisted in this task by the domestic production of crude oil which insulated the country from the immediate effects of the energy crisis. Although imports of crude oil were growing steadily, these were principally directed towards the 'refining for export' and petrochemical industries that Romania was expanding throughout the decade and the value of crude imports was offset by exports of oil products.

A series of events interrupted this pattern in 1977. In March a major earthquake caused considerable damage to housing and factories in the country's major industrial regions, including Bucharest, and also damaged oil installations. Ceausescu responded to this disaster by announcing increases in the level of investment and industrial output which were to be achieved by increased labour productivity and reduced expenditure on raw materials. Simultaneously, increases in wage rates and social security payments were announced (*Scinteia*, 27 June 1977).

It is difficult to tell how much these increases should be attributed to rhetoric rather than intent. In May 1977 a series of changes in the method of taxing income and expenditure were announced, the net effects of which were that a new tax was imposed on enterprise profits that was partially offset by a reduction in turnover tax, while the increased revenue was channelled into investment in heavy industry (Smith, 1982b). Taxes on individual incomes were replaced by a tax on the enterprise wage fund and paid at source. Wage rates have subsequently been announced net of tax rather than gross (a reduction of approximately 13 per cent), making it difficult to compare actual wage payments before and after the new system was imposed. (Published data for all years back to 1950 have now been recalculated on the new basis in statistical publications since 1981, for periods when the system was not actually operating.)

There is, however, a broad consensus from Romanian and CMEA sources that average wages grew by more than 40 per cent over the five-year plan period, although the annual breakdown of this growth is subject to some dispute. Data provided to the IMF do not coincide with official statistics, and growth rates in Table 7.2 have been measured gross up to 1976, and net thereafter.

This growth of money wage rates amounted to a wage explosion by Romanian standards (wage rates growing by half as fast again as in each of the three preceding five-year plans) as a result of which total wage payments grew 16 per cent faster than planned state retail turnover over the five-year period.

Although the state retail turnover plan for 1980 was fulfilled in money terms (and was in fact the only supply target established in the five-year plan to be reached) this was only achieved by the overfulfilment of car sales, while planned targets for the sales of fruit, vegetables, eggs, footwear, radio sets, televisions and refrigerators were all substantially unfulfilled. Actual levels of sales were also assisted by a growth of imports or a reduction in exports. Domestic output of meat, refrigerators, radios and TV sets in 1980 were either below or barely exceeded 1976 levels, while car production was underfulfilled by 23 per cent (23,000 cars) so that above plan sales were only achieved by substantial imports. The plan communiqué for 1980 also referred specifically to giving up quantities of exports in order to maintain consumption.

The wage explosion appears to have been triggered off by attempts to 'buy off' popular discontent, following the earthquake and the overambitious demands made by the Party from the workforce in the face of the debilitating effects of national disaster.

Labour unrest has been admitted obliquely in official statements as a cause of low productivity and has resulted in open confrontations, the most serious of which have been reported in the Jiu Valley, the country's principal mining region in 1977, 1979 and 1981. Although the disturbances were partly motivated by social and political factors, the major cause appears to have been dissatisfaction with living standards, work conditions and remuneration. The Romanian government responded by making monetary concessions, including increasing the wage rates of industrial workers by 11.3 per cent in 1978, as a result of which *net* wages were restored to a level of 5 per cent higher than 1976 *gross* wage rates, and savings deposits grew by 24 per cent. The real increases in reported real turnover were largely assisted by a drastic reduction in trade surpluses in foodstuffs from their (record) 1977 levels which, combined with the rising crude-oil bill, has resulted in substantial balance of payments deficits, totalling over one billion dollars per annum since 1978.

In February 1982, official increases in food prices of 35 per cent were announced (*Scinteia*, 10 February 1982). The effect of food price increases was officially estimated to raise the cost of living by 14.8 per cent and was accompanied by an announcement of wage increases of 16.5 per cent (over 1980 levels), increases in pensions of 15.7 per cent and child allowances of 35 per cent (to compensate for the full impact of food price rises). Wage and pension increases were progressive with no increase given to those in the highest wage and pension categories (*Scinteia*, 10 to 16 February inclusive). The policy appears to be intended to preserve the purchasing power of low-income groups, while forcing high-income groups to cut back savings. Unofficial western estimates, based on observations of prices in state stores, indicate that food prices may have risen in reality by 70 per cent. Subsequently, petrol prices have also increased.

If these observations are confirmed, state retail prices will have risen by 25 to 30 per cent, which would involve a reduction in real wages of 10 per cent, not the officially stated increase of 1.5 per cent.

Bulgaria

In Bulgaria, the supply of labour to the state sector continued to grow at an average of 6 per cent per annum until 1976. Attempts to reduce the rate of growth of wage rates were only partially successful and the rate of growth of retail turnover of 8 per cent was facilitated by rising indebtedness. Wage rate growth slowed down in the mid-1970s, but in 1977 labour supply to industry fell and real income per head grew by only 0.6 per cent. From 1978 onwards wage rate growth has accelerated with no corresponding growth in the volume of real retail turnover, again resulting in open inflation. A reform of wholesale prices was announced in January 1980 which was intended to bring domestic prices into line with world market prices (Kaser, 1981) and give producers an incentive to respond to world market conditions. As a result retail prices increased by 13.9 per cent but wage rates were simultaneously increased by 17.0 per cent and the real volume of retail turnover grew by 3.6 per cent.

Hungary

The removal of direct central controls over wages and prices and the virtual abolition of the *Preisausgleich* was an integral feature of the New Economic Mechanism. Until 1968 Hungarian practice in inflation control was very similar to that of other CMEA countries. Initial attempts to enhance the role of piece rates in wage determination from 1950 to 1956 had been superseded by the traditional system, in which changes in wages were linked to changes in gross output (or productivity measured in terms of gross output), subject to a central limit on the overall wage fund, determined by the Central Planning Office, which was then disaggregated to branch ministries and thence to enterprises (Marrese, 1981, pp. 63–4).

The principal intention of the NEM was to make enterprises more responsive to changes in domestic and international market conditions, and was to involve the replacement of detailed instructions to enterprises by economic levers. This process was to involve the discontinuation of annual operational plans and the central fixing of wage scales, and the virtual abolition of the *Preisausgleich*. Wholesale prices were to be made more rational by the introduction of capital charges, and by the inclusion of permanent changes in world market prices.

Although enterprises were to have greater freedom in determining output prices, including those for consumer goods, it is not surprising, in view of the monopolistic structure of Hungarian industry, that central authorities retained the power to affect enterprise profitability by altering taxes and subsidies and so on, if such price increases were due to factors other than cost changes and involved elements of profiteering. As it was anticipated that retail prices would in general be 6–10 per cent higher than producers' prices, changes in foreign trade prices that affected wholesale prices would in turn affect retail prices (Marer, 1981, p. 169).

Changes in wage payments were to be financed out of enterprise profits. After all compulsory payments, including taxes, contributions to reserves and repayments of credit obligations had been met, the enterprise was to be free to distribute remaining profits between a sharing fund for workers and investment. In order to encourage ploughback and discourage excessive wage payments the sharing fund was subject to progressive taxation after certain levels had been met (see below).

Central planners were therefore to dispense with many of the direct controls over wages and retail prices which had formed the mainstay of the system of inflation control in centrally-planned economies. In view of the experience of hyperinflation in the immediate post-war period some Hungarian economists felt that the population was so conscious of inflationary pressures that it overestimated their impact on living standards. As a result, the control of inflation was considered to be vital to the success of the reforms to the extent that it was even argued that the reform mechanism would be jeopardised if the cost of living rose faster than 3 per cent per annum (Granick, 1975, p. 251).

This fear of the consequence of inflation, combined with the desire to preserve justice in income distribution and to maintain full employment, has required central authorities to use their tax and subsidy powers and to intervene with direct controls, to the extent that these have frequently conflicted with the objectives of economic efficiency that the reforms were intended to stimulate.

For example, microeconomic efficiency requires that increases in the real cost of imported commodities or of the cost of production of domestically produced commodities should be passed on to final consumers in the form of changes in retail prices. In theory, changes in the relative prices of imported or domestically produced commodities *need* not lead to changes in the absolute price level, provided a strict control over aggregate demand (and money supply) is maintained, which will mean that increased prices in one sector will be offset by price reductions elsewhere. Those working in activities that have become relatively more costly, and/or less demanded will find their profits and incomes reduced, while those working in competing fields will find their profits and incomes increased. In theory, therefore, workers should transfer out of the activities in reduced demand to those in increased demand.

Hungarian economists have recognised that in practice this process is far from instantaneous, particularly in a highly-monopolised industrial structure, in which vested interests may successfully resist reductions in money income and pressure central authorities to grant subsidies, or relax the terms of credit repayments or tax payments in order to preserve profit levels (Kozma, 1981). Furthermore, as Granick has shown (1975, pp. 245–50), the fear of having to render uneconomic plants, or even whole enterprises, bankrupt and create unemployment has resulted in central authorities willingly conniving at this process. Granick proposed that the best policy to preserve full employment while simultaneously avoiding open inflation is to set prices at a level that will cover the costs of less efficient producers, but maintain a volume of excess demand so that all output can be sold at those prices. The implications of this policy are that the state must be involved in price-fixing in retail markets, while equity requires that the excess profits of efficient producers (ie, those using more economic processes) should be heavily taxed. Thus the preservation of macroeconomic goals severely constrains microeconomic efficiency.

Although Portes and Winter (1980) have argued that the Hungarian economy was not characterised by excess demand in the period 1968–75, Granick's predictions of the problems Hungarian authorities would encounter and the policies they would pursue to combat them have proved to be remarkably accurate in the period of rising energy prices.

The authorities have found themselves in a position where, on social grounds, they are effectively forced to guarantee minimum wage increases while imposing a ceiling above which wage increases will be progressively taxed. Social justice also requires that this ceiling must be imposed on the enterprise *not* the individual (otherwise individuals receiving high percentage increases on low incomes would be subject to higher taxation than those receiving low percentage increases on high incomes). Enterprises then distribute the post-tax sharing fund between workers as they choose. This process requires the formulation of definitions of the amount of payments than can be made out of the profit-sharing fund, before taxation will be imposed.

Two basic systems have been utilised for calculating the size the profit-sharing fund may reach before being subject to taxation; the first linked to the average wage rates in a base period, and the second to the total wage fund in the base period (Marrese, 1981).

The untaxed profit-sharing fund of enterprises in the first system comprises a centrally-determined guaranteed minimum increase, calculated as a percentage of the average wage rate, plus an indicator linked to enterprise profits up to a centrally determined ceiling, calculated as a percentage growth of the average wage rate. Enterprises in the second system can pay a centrally-determined, guaranteed minimum increase calculated as a percentage of the total wage fund, plus an indicator linked to value-added, up to a centrally-determined ceiling again calculated as a percentage growth of the average wage rate. After the ceiling is reached the profit-sharing fund is taxed progressively. From 1976 to 1978 the guaranteed increase was 1.5 per cent and the ceiling 6 per cent for both systems, considerably limiting the incentive effects of the profit or value-added indicators.

As the base indices on which the size of the profit-sharing fund is calculated refer to a preceding period, not the current period, the system linked to the average wage rates is designed to allow enterprises which are expanding their labour force to increase the size of the fund without incurring taxation as the labour force expands, while the system linked to the total volume of wage payments provides an incentive to enterprises to decrease their labour force by permitting them to distribute the same level of profits amongst a reduced labour force (up to a fixed per-capita maximum) before incurring taxation.

In the initial period of the reform 93 per cent of enterprises operated on the wage rate system of regulation. Enterprises soon discovered that higher wage increases could be paid to the existing labour force by employing more labour at wages below the enterprise average (Marrese, 1981, p. 66). This could of course have stimulated total factor productivity by encouraging workers to move from below average productivity enterprises to above average productivity enterprises. This however presupposes that the *marginal* productivity of the workers so transferred would be higher in the new enterprises than in the old. In practice the rate of growth of labour productivity in state-owned industry declined from 5.7 per cent per annum from 1961 to 1967 to 3.2 per cent from 1968 to 1971 (Granick, 1975, p. 306) a decline which Marrese

(1981, p. 66) attributes to the employment of increased volumes of low-productivity labour as a result of the wage rate system of calculating profits. The real volume of retail trade was increased (assisted by improvements in the terms of trade) and open inflation remained low in this period. There has subsequently been a continuous movement away from wage rate regulation towards regulation based on the total wage fund which accounted for 55 per cent of all enterprises in 1978 while a further 30 per cent of enterprises do not retain indicators linked to productivity and are taxed on any increase on average wages in excess of 6 per cent (Marrese, p. 71).

In the period since the increase in world energy prices, which has resulted in a substantial deterioration in the terms of trade, the problem of controlling inflation and stimulating an efficient allocation of resources has provided a far greater dilemma for Hungarian authorities, which they alone in the bloc have attempted to cope with without the immediate benefits of a *Preisausgleich*, but with an exchange rate system which is intended to feed foreign trade prices directly into enterprise accounts.

The Hungarian exchange rate is estimated on the basis of the average cost of production of actual exports, measured in domestic wholesale prices, divided by the foreign exchange actually obtained for the sale of those items (Marer, 1981, p. 168) (separate rates have been used for transactions in convertible currencies and transactions in transferable roubles throughout the 1970s). The use of an average rate rather than a marginal rate (ie, the ratio of the domestic cost to foreign exchange earnings of the least profitable item to be exported) results in an overvaluation of the exchange rate which may involve financial losses to exporters with above average costs and makes imports less costly, thereby keeping down the level of wholesale and retail prices, but making imports more attractive to domestic users.

In the period preceding the energy crisis, subsidies had to be paid to exporters to stimulate a sufficient volume of hard currency earnings to pay for imports, which resurrected the process whereby enterprise managers stood to gain as much from the process of bargaining with superior authorities as they did from making genuine economic substitutions (Marer, 1981). The effect of the energy crisis was that import prices rose faster than export prices in general, while prices of some imports and exports rose substantially faster than others.

One method of coping with foreign price changes would have

been to revalue the currency and allow changed relative prices to feed directly into the domestic price structure. If the revaluations had been based on the rate of change of import prices the domestic price level would have remained unaffected in the aggregate, with some prices increasing and others being reduced. This policy would however have required substantial increases in export subsidies to preserve external equilibrium which would in turn require increases in taxation elsewhere. Under these circumstances the process of stimulating adaptation by price criteria would have resulted in windfall gains for some sectors of the community and losses for others, while the economic requirement to pay out subsidies to exporters while changing tax rates would have left central authorities open to special pleading, and would have made it impossible for them to appear 'neutral' in the adjustment process (Kozma, 1981).

It is not surprising that the authorities gave greater emphasis to domestic equilibrium, leaving exchange rates virtually unaltered and attempting to stabilise the domestic price level. As a result, rising world market prices meant exports became more profitable and export subsidies on specific items were reduced (or replaced by taxes) while imports had to be subsidised. The policy of specific subsidies and taxes on imports and exports which prevented domestic prices from reflecting changes in world market prices, implied a *de facto* restoration of the *Preisausgleich*, gave domestic consumers no incentive to cut back on relatively expensive items, and placed much of the burden of adjustment on the foreign trade sector.

In 1976, as the impact of Soviet oil price increases started to be felt, the forint was revalued against both the dollar and the transferable rouble (Kozma, 1981, pp. 213–14), import subsidies were reduced and retail prices rose by 5.0 per cent. Real wages and real incomes grew by 0.1 per cent and 0.3 per cent respectively (Table 7.2), the lowest increase for 20 years. In 1977 wage rates grew by 8.9 per cent as workers obtained increases in money rates that would compensate for the preceding years' inflation and leave a real increase of around 3 per cent if the retail price inflation were to continue at the same rate. Subsidies were broadly retained in 1977, as a result of which the retail prices of many commodities, particularly petrol and domestic heating oil, were below the wholesale price, while the subsidy for foodstuffs amounted to 26 per cent of the retail price (Marer, 1981, p. 168). The volume of real retail turnover grew by 6.2 per cent, largely as a result of a doubling of the trade deficit with the West to 1.6 billion dollars, permitting an increase in real wages and real incomes. This pattern, although clearly not sustainable in the long run was broadly repeated in 1978.

Starting in 1979, the Hungarian authorities reconfirmed their intention to tackle the underlying problems of macroeconomic equilibrium by the methods of the New Economic Mechanism rather than by continued centralisation. Price increases were announced that restored the principle that retail prices should exceed wholesale prices, and removed many of the subsidies on consumer goods (Marer, 1981; Tyson, 1981). Open inflation amounted to 8.9 per cent and the announced increases in wages and other incomes did not compensate, resulting in a fall in real wages by 1.7 per cent and real incomes by 0.3 per cent. Further reforms in wholesale prices were announced in January 1980 bringing prices more in line with foreign trade prices, simultaneously increasing prices and reducing the need for specific subsidies (Marer, 1981, p. 196; Hare *et al.*, p. 20). Preliminary estimates indicate that this move has resulted in further decreases in real income in an attempt to restore internal and external microeconomic and macroeconomic equilibrium.

PART THREE

INTERNATIONAL ECONOMIC RELATIONS

8 THE ECONOMIC AND INSTITUTIONAL BACKGROUND TO EAST EUROPEAN TRADE

The combination of economic and political factors which Stalin claimed had led to the development of two competing world economic systems and the emergence of competing economic blocs within a Europe divided by borders dictated more by political treaty than by economic or geographic rationality, required the East European economies both to redirect their trade flows and to fundamentally reorganise their foreign trade systems.

The Redirection of Trade

Two separate factors contributed to the redirection of trade flows. In the inter-war period the East European economies had conducted the vast majority of their trade with West European countries and with Germany in particular. The western economies were reluctant to fill the vacuum left by the destruction of the German economy and, in particular, were unwilling to grant credits to help in the reconstruction of the economies of the former belligerent economies of Hungary, Bulgaria and Romania, while questions of reparations and pre-war liabilities were outstanding (Berend, 1971).

East European participation in the Marshall Plan for European Reconstruction based on substantial quantities of American capital aid (announced on 5 June 1947 before the East European countries had become fully integrated into the Soviet bloc) was effectively vetoed by Stalin himself to the detriment of Czechoslovakia who had agreed to participate and Hungary and Poland who had expressed interest in participation. Stalin's objections were probably based on his desire to preserve his influence over economic and political development in Eastern Europe and on his personal predilection for the dominance of steel production which would have been threatened by the subsequent proposals for dollar viability which caused dissent when Greece and Norway proposed to construct steel plants (Kaser, 1967, p. 19).

The prospects for the re-establishment of trade relations between Eastern and Western Europe declined further during the period before Stalin's death. The American embargo on exports of strategic and technological products, which accounted for approximately 50 per cent of internationally traded products to Eastern Europe, was extended to prevent all recipients of Marshall Aid from trading with Eastern Europe, a policy which was co-ordinated by the establishment of CoCom (van Brabant, 1980, p. 30). The subsequent slide into cold war, the Berlin blockade and the Korean War meant that by the time of Stalin's death in 1953 the moves towards economic union and integration in Western Europe were sufficiently advanced to prevent the re-establishment of former or potential trade links.

The development of West European integration was initially most damaging to the more industrialised North Eastern and Central Eastern countries, who in addition to suffering from the effects of the strategic embargo (a factor which hindered Czechoslovak growth) also had their export prospects damaged by the development of the European Coal and Steel Community (which particularly affected Poland). Later the agricultural policies pursued by the EEC were to have similar effects on the export prospects for the agricultural outputs of Romania, Bulgaria and Hungary.

The principal effect of isolation from trade with the industrial economies of the West was to prevent the East European economies from importing higher-quality machinery and equipment and consumer goods, and to prevent them from earning the hard currency required to import basic fuels and raw materials. The basic impact on their domestic industries was that they were unable to benefit from the economies of scale arising from the exploitation of larger markets, while the isolation from foreign competition removed much of the incentive for domestic industries to improve their quality specifications.

The Soviet Union was only partially capable of filling these gaps. Although Soviet resource endowments meant that it had the physical reserves to meet East European demand for most basic materials, the location of the major natural deposits (particularly in the harsh terrain of western Siberia) meant that the costs of prospecting and extraction were high by world standards, while the transportation of raw materials across an enormous land mass posed severe economic and physical problems. The Soviet Union was less suited to the supply of high-quality machinery and equipment and consumer goods, while the conditions of a sellers' market meant that consumers and enterprises were not particularly demanding about the quality of the commodities they received.

Some scope for the solution of these problems might have been found with improved integration *between* the East European countries themselves. Clear areas of complementarity arose between the industrial economy of Czechoslovakia, the coal-producing regions of Poland and the agrarian and textile economies of Hungary. Although in the immediate aftermath of the war the East European countries had little choice other than to trade with the Soviet Union, particularly for the supply of grain and raw materials, by 1948 their mutual trade had overtaken their trade with the USSR. Trade between the Central East European nations had grown to such an extent that by 1948 only 10–20 per cent of their trade was conducted with the USSR, while the Balkan nations had also been drawn into the process with Bulgaria and Romania conducting 75 per cent and 50 per cent respectively, of their trade with other East European countries (Berend, 1971, p. 10).

Of greater significance were the number of economic agreements signed between the East European countries, which in addition to co-ordinating their plans for economic reconstruction, incorporated proposals for customs unions and therefore implied the continued use of market levers for their trade.

Yugoslavia and Albania were the first to sign a thirty-year agreement for 'co-ordination of economic plans, union of customs areas and standardisation of currency' in December 1946 (Berend, 1971, p. 12). The Czechoslovak-Polish agreement signed in July 1947 established a permanent joint economic committee which drafted specialisation agreements to limit investments to those industries in which the countries had a comparative advantage. Similar agreements were signed in 1947 and 1948 between Hungary and Yugoslavia, Romania and Bulgaria, and were proposed between Hungary and Czechoslovakia. The Romanian-Bulgarian agreement explicitly mentioned the development of a customs union, while the President of the Hungarian Planning Office announced that the purpose of the agreements was 'to adjust the development of industrial branches . . . [so that] . . . a more or less coherent economic system should be developed by the Eastern European People's Democracies' (Berend, 1971, p. 12). Hungarian authorities have made it clear that these proposals were vetoed by Stalin himself, and the *radial* pattern of trade emerged in which the

East European economies conducted the vast majority of their trade with the Soviet Union while remaining fundamentally unintegrated between themselves (Berend, 1971).

The process was partly caused by the policy of according priority to heavy industry espoused by Stalin, and by the exigencies of the Korean War which required each country of the bloc to increase its output of iron, steel, heavy engineering products and electricity (and where possible coal) as rapidly as possible. This inevitably raised the demand for energy and raw materials in the bloc and required the resource-poor countries to import from the USSR. Although the Central East European countries still conducted a considerable volume of trade amongst themselves, 25–30 per cent of the reported trade of Hungary, Czechoslovakia and Poland was directed towards the USSR in the early 1950s. The effects of the 'radial' pattern were more apparent in the case of the Balkan countries. In addition to reparation payments, over half of Bulgaria's trade and possibly 75 per cent of Romania's trade was concentrated with the USSR. As a result, the USSR imported machinery and equipment and other engineering and steel products from Czechoslovakia, Poland, the GDR and to a lesser extent Hungary, and exported crude oil and oil products, coal, iron ore, ferrous and non-ferrous metals and ores, grain and cotton to those countries. Simultaneously, the USSR exported machinery and equipment, engineering and steel products to Bulgaria and Romania and imported ores and metals and agricultural products from Bulgaria, oil and oil products and grain from Romania and also imported ferrous and non-ferrous metallic ores from Czechoslovakia.

The Development of the Soviet Foreign Trade System

The Soviet system of administration of foreign trade was originally developed to cope with the organisation of trade between a socialist economy and capitalist economies, rather than between centrally-planned economies themselves. The adoption of central planning in Eastern Europe therefore necessitated changes in the methods of conducting foreign trade in both the USSR and Eastern Europe.

The principal feature of Soviet administration is that the state (through the aegis of the Ministry of Foreign Trade) has the exclusive right to conduct foreign trade transactions. Individuals and enterprises are therefore prevented from directly purchasing or selling commodities in foreign markets.

During NEP these rights were largely exercised by individual trade delegations operating *outside* the USSR who operated in harness with orders placed with and by domestic state trading organisations under the supervision of the Commissariat of Foreign and Domestic Trade (Quigley, 1974, pp. 45–60). In addition, other state-owned organisations (including enterprises) and co-operatives and mixed companies (involving foreign ownership) were permitted to engage in foreign trade, provided such trade was limited to products of their own manufacture or for their own production requirements (Baykov, 1946, p. 12). A further feature of NEP was the attempt to attract foreign capital by granting overseas companies the right to operate concessions in the development of raw materials (mainly oil and timber). The policy was largely unsuccessful and in 1928 concessions accounted for only 0.6 per cent of industrial output (Nove, 1969, p. 89).

A number of changes were implemented during the first five-year plan which laid the foundations of the current system which was subsequently installed in Eastern Europe. Domestic and foreign trade operations were separated and no agencies were permitted to engage directly in foreign trade other than the Commissariat (subsequently the Ministry) of Foreign Trade which conducted all foreign trade operations through combines specialising in particular commodity groups. Finally it was decreed in 1935 that foreign business should be transacted within the USSR not outside it, considerably strengthening the bargaining position of the Commissariat with recession-hit foreign manufacturers (Baykov, 1946, p. 21).

The Stalinist foreign trade system was specifically designed to enable extensive growth to take place against the background of western recession. The major benefits of the system were that it protected new industries from foreign competition, allowed available foreign currency earnings to be allocated according to centrally-determined priorities and eliminated competition between Soviet purchasing and sales agencies on foreign markets.

The second outstanding feature of Soviet foreign trade in the pre-war era was its low volume in relation to GNP. This is in part determined by country size (the volume of foreign trade is normally inversely related to population size and land mass) and by the abundance of natural resources which reduce the physical requirements for imported raw materials. It is often proposed that the USSR participates in trade to a lesser extent than would be determined by these factors alone, and that this is derived from a desire to remain as self-sufficient as possible (autarky) even though this policy may result in economic costs (eg, the cost of obtaining and transporting raw materials from Siberia might be more than the domestic cost of producing the exports required to import these materials).

Autarky was not part of the Bolsheviks' original conceptions, particularly as far as trade with other socialist states was concerned. Bukharin and Preobrazhensky (1970, p. 326) lamented the costs incurred by the destruction of pre-war trade relations and urged that they should be re-opened as far as this was consistent with the state's general aims, and that this would be facilitated by revolutions in other countries. Preobrazhensky's policy of according priority to heavy industry did involve considerable elements of import substitution, whereas Bukharin's policies would have involved a greater volume of trade along the lines determined by comparative advantage. The principal effects of the pursuit of Preobrazhensky's policies should have been therefore to affect the structure and direction of foreign trade once other socialist states had been developed rather than its absolute volume.

In practice, the low levels of foreign trade pursued by the Soviet Union in the 1930s were as much a product of external circumstances as design. Foreign trade was planned to grow 2½ times in real terms during the course of the first five-year plan. By 1938, Soviet foreign trade was only 33 per cent of its 1929 level in money terms and approximately 66 per cent of that level in real terms.

The contraction was largely the result of the world recession which, according to estimates based on the Soviet physical volume for trade, significantly worsened Soviet terms of trade during the first five-year plan, and reduced the profitability of foreign trade to the Soviet economy by reducing world demand for staple Soviet exports (foodstuffs and raw materials). Trade turnover reached an inter-war peak in money terms in 1930 and in real terms in 1931, before halving in real terms in the second five-year plan. The terms of trade moved so sharply against the USSR in 1932 that a potential surplus in 1929 prices became a substantial deficit. Imports declined in real and money terms in 1933, while the need to pay back foreign credits required exports to decline more slowly from 1933 to 1935 until the trade deficits accumulated in the first five-year plan had been covered by surpluses.[1]

Problems of the Adoption of the Foreign Trade Monopoly in Eastern Europe

The major advantage of the foreign trade monopoly in the USSR in the 1930s was to strengthen the bargaining position of the Soviet state with capitalist enterprises and to ensure that foreign exchange was concentrated on the commodities that were considered to be most important for economic development. The imposition of the Stalinist system of central planning in Eastern Europe clearly called for some revision of the foreign trade system and, in particular, for a differentiation between the methods used for trading with other socialist states and with capitalist nations. The imitation of the Soviet planning system in each of the East European countries, each with separate centralised foreign trade systems, does not offer the East European states the most appropriate mechanism either for conducting their mutual trade or for conducting trade with capitalist nations.

The structure of foreign trade in a centrally-planned economy is largely determined by the system of planning by material balances. The process involves drawing up an initial production plan from which an import plan can be derived and then redrafting both the production and import plans to take account of the exports required to pay for imports (Boltho, 1971, pp. 54–5).

Imports are therefore basically restricted to items that either cannot be produced domestically in the quantities required in the plan period or at sufficiently low cost. Exports largely consist of commodities for which domestic production exceeds demand either as a result of resource endowment (Soviet minerals and energy), a deliberate investment policy or even a past surge in production for which the domestic market is now saturated.

The centralised planning of supply prevents an enterprise from responding to orders arising from outside the planning system; the principle is implemented, as far as domestic transactions are concerned, by passive money. The extension of the principle to the foreign trade sector prevents an enterprise from bypassing its own central planning authorities and either purchasing inputs or selling outputs directly from/to enterprises in another country. Before trade can take place orders must be directed through the foreign trade ministry and the internal production and supply arrangements made by the appropriate domestic central planning authorities. Under these circumstances the national currencies of the

East European countries are said to be 'commodity inconvertible' — the foreign owner (western or eastern) of an East European currency cannot convert it into any commodity he may demand (eg, the owner of Romanian leu cannot demand payment in oil unless a corresponding plan instruction has been made).

A further factor that hinders trade flows is that domestic currencies are not financially convertible either — the owner of an East European currency cannot demand that it be converted into other currencies (eg, dollars) on demand at the official exchange rate.[2]

The establishment of separate foreign trade monopolies has prevented the East European nations from providing a united front in their relations with the West leading on occasions to direct competition. In the automobile field, for example, East European nations have entered separate negotiations for the purchase of licences for production of cars, lorries, buses and associated spare parts as a result of which different East European countries have brought licences to produce very similar products from different western manufacturers (Pecsi, 1981, p. 44).

Furthermore, as a result of the protectionism engendered by the foreign trade monopoly East European enterprises have been isolated from innovations and technical developments occurring in the West, while the fact that they do not compete directly in overseas markets means that they are frequently slower to anticipate and take advantage of changing western market demands.

The disadvantages of separate centralised foreign trade monopolies for each of the centrally-planned economies are more acute in their mutual trade. In precisely the area where economic relations should be based on co-operation and planning, separate foreign trade ministries representing national interests confront each other in a position of bilateral monopoly. As a result, intra-CMEA trade negotiations are conducted bilaterally between the foreign trade ministries of the socialist states who agree in the first instance on trade protocols over a five-year period coinciding with the time span of the individual five-year plans. The five-year trade agreements are highly aggregated and are largely confined to specifying the value of mutual trade, the broad commodity groups to be included and certain more detailed specifications concerning the delivery of specific items. The five-year agreements are subsequently broken down into annual agreements which will contain a far greater degree of detail concerning, for example, product specifications. Finally, specific questions concerning the timing of deliveries, quality, and so on, may be left to be negotiated between supplying and receiving enterprises. Deliveries agreed in the five-year and annual protocols should be considered as basic minima, the sum of the annual protocols may exceed targets established in the five-year plans, while deliveries over and above those established in annual protocols may be made during the course of the year.

Multilateralism and Bilateralism

The problem of financial inconvertibility in East-West trade is largely overcome by conducting that trade in western (convertible) currencies. This follows the Soviet practice of the 1930s and allows the ministries of foreign trade to seek out the best market conditions for the purchase of imports and sale of exports independently of one another.

Soviet and East European trade with the industrial West has therefore tended to be multilateral in character. Various tests have been devised to measure the degree of multilateral balance in visible trade. The simplest of these, used by Pryor (1963), sums the value of visible trade surpluses and deficits with each individual country and expresses this as a percentage of total trade turnover. This is not strictly speaking an index of multilateral balance as it does not indicate whether the volume of trade in total was balanced. A variant originally devised by the League of Nations (1933) to deal with this problem subtracts the volume of unbalanced trade from both the numerator and the denominator in Pryor's index, and provides an indication of the amount of balanced trade that was balanced multilaterally.

The indices are given as follows:

$$\text{(1) Pryor's Index} = \sum_{i=1}^{n} \frac{Xi - Mi}{X + M}$$

$$\text{(2) League of Nations Index} = \sum_{i=1}^{n} \frac{|Xi - Mi| - |X - M|}{X + M - |X - M|}$$

Where Xi = exports to country i
Mi = imports from country i
X = total exports to countries observed
M = total imports from countries observed

The tests are subject to a number of statistical limitations including the method of measuring trade and the method and comprehensiveness of partner identification. The broad conclusions are not substantially affected by allowances made for these factors, particularly for those countries for whom sales of arms and precious metals are insignificant.

An analysis of the Pryor and League of Nations indices estimated for each East European country from 1960 to 1973 (the last year before the energy crisis) indicates that (1) between 18 and 25 per cent of trade with market economies is not balanced bilaterally (Pryor's Index); and (2) between 14 and 20 per cent of trade with market economies is balanced multilaterally (League of Nations Index).

Table 8.1 provides a comparison with indices calculated for France and West Germany. (The years 1961, 1963 and 1969 were chosen to avoid the effects of western exchange crises.) The table indicates that there is no clear difference between the degree of non-bilateral and multilateral trade with market economies shown by East and West European countries, and that in the period before the energy crisis, the CMEA nations did not differ significantly in the degree of multilateralism in their trade with capitalist nations from the capitalist nations themselves.

A comparison of indices for the 1930s indicates that multilateral balancing was a feature of Soviet, not East European trade. The German pursuit of bilateralism under Schacht in the 1930s forced the East European countries to reduce the volume of multilaterally balanced trade to around 10 per cent. The USSR was also forced into bilateralism in 1934 and 1935 due to the need to reduce its deficits with Germany. Although the absolute volume of trade was substantially reduced in the remainder of the decade the USSR used the trade monopoly to pursue a multilateral policy exporting wood, grain and lumber to the United Kingdom and importing machinery and equipment from Germany and the USA.

This multilateralism does not extend to intra-CMEA trade. The absence of financial and commodity convertibility provides no incentive for one country to accept another East European currency

Table 8.1: Indices of Multilateral Trade With Market Economies

	1961 Index		1963 Index		1969 Index	
	Pryor	LON[a]	Pryor	LON	Pryor	LON
Romania	19.9	14.8	21.0	13.6	23.9	12.9
Bulgaria	25.0	23.1	18.8	14.4	16.9	15.0
Poland	22.0	17.3	20.3	20.0	18.1	17.4
Hungary	18.7	10.7	20.0	14.5	16.9	14.0
Czechoslovakia	18.9	17.2	17.8	15.5	18.9	14.2
France	21.0	19.1	23.6	15.3	13.4	10.4
West Germany	20.7	15.3	19.9	18.8	17.5	7.4

Note: a. League of Nations.

Source: Calculated from the statistical handbooks of the countries concerned.

as a means of payment. As a result, in the early 1950s each CMEA country attempted to balance its trade with other member countries on a strict bilateral basis. Clearing accounts were opened in the State Banks of the respective countries and accidental imbalances in the mutual trade of any two countries in one year were offset by a compensating imbalance in a subsequent year. A preliminary attempt to create a multilateral clearing system was made in 1957. Payments for deliveries in excess of those established in annual protocols could be paid into accounts held at the USSR State Bank, and, subject to the agreement of all parties concerned, surpluses and deficits could be cleared multilaterally. Outstanding debts had to be paid by deliveries of commodities detailed in a specific list (Schiavone, 1981, pp. 140–1).

The current system involving the establishment of an international clearing bank — The International Bank for Economic Co-operation (IBEC) — and an international unit of account — the transferable rouble (to which the domestic currencies are linked) — was instituted in 1964.

Member countries' accounts with the Bank are held in transferable roubles and are credited and debited by the delivery of invoices to the Bank by exporting countries who simultaneously invoice the importing country (Schiavone, 1981, p. 159). In theory, therefore, debits and credits could be cleared on a multilateral basis without the need for bilateral balancing and it was hoped that a trade surplus arising between (say) Czechoslovakia and Bulgaria would be offset by deficits arising between (say) Hungary and Bulgaria and between Czechoslovakia and Hungary.

It was not anticipated that these surpluses and deficits would arise in a totally haphazard fashion — but could be planned in advance by the member countries. Member countries were charged that: 'when concluding commercial agreements . . . to ensure the balancing of receipts in transferable roubles as a whole with all other member countries of the bank within any one calendar year . . . or period agreed on' (CMEA, 1977, p. 291).

This did not necessarily imply a strict balance in trade and commercial payments as credit operations were to be included as balancing items.

The crux was that the transferable rouble was ultimately to be financially convertible; Article II(a) of the Bank's current charter states: 'The Bank Council shall study the possibility of the Bank's carrying out operations for the exchange of transferable roubles

Table 8.2: Multilateral Balance in Intra-CMEA Trade, 1961–70 (excluding Mongolia)

	1961–5		1966–70	
	(a)	(b)	(a)	(b)
Bulgaria	1.9	43.1	0.9	70.0
Czechoslovakia	1.5	19.1	1.3	37.9
Hungary	1.9	31.4	1.8	36.9
Poland	2.0	16.1	2.0	21.0
Romania	0.3	65.5	0.3	44.4
Average	1.5		1.5	

(a) League of Nations Index.
(b) Percentage of annual imbalances corrected bilaterally (see p. 161).
Source: Calculated from statistical yearbooks of countries concerned.

for gold and other currencies' (CMEA, 1977, p. 292).

The original charter proposed that the study would be implemented 'within one year of the Bank's foundation' (Ausch, 1972, p. 159). As a result, therefore, the member countries would be compensated for trade surpluses arising in their mutual trade. The implementation of the proposals foundered on the fact that they offered too *great* an incentive to establish trade surpluses in intra-CMEA trade and a consequent disincentive to establish deficits.

The lack of success of these measures may be judged by the League of Nations indices estimated for the five-year plan periods for 1961–5 and 1966–70 shown in Table 8.2. A five-year period has been chosen (ie, all measurements of exports and imports have been aggregated over a five-year period rather than a single year) in order to eliminate the effect of an accidental surplus, or deficit, in trade with any country in one year being offset by a deliberate deficit or surplus with the same country in a subsequent year (the practice which the clearing rouble and the transferable rouble were specifically designed to eliminate). This is supplemented by column (b) which indicates the proportion of annual imbalances with partners which have subsequently been offset in this fashion (ie, a zero figure would indicate consistent surpluses and deficits in trade with each individual trade partner and a figure of 100 would indicate that annual surplus/deficits with each trade partner have been offset by compensating surpluses and deficits with that partner within the five-year plan period).

Table 8.2 shows that only 1.5 per cent of the mutual trade of the

East European countries and the USSR is cleared multilaterally in each five-year plan period, and that the volume is unchanged in the plan period immediately following the establishment of IBEC and the transferable rouble. It could be argued that the observed reciprocity could be caused by purely random factors and the comparable figure for intra-EEC trade from 1961 to 1965 is only 3 per cent.[3] This can be mainly attributed to the fact that surpluses in intra-EEC trade are paid for in hard currencies that can be used outside the bloc — there is therefore no imperative to balance intra-bloc trade in the aggregate which is reflected in the low degree of multilaterally balanced trade. The consistency with which trade between any two countries can be characterised by a persistent surplus or deficit is noticeable in intra-EEC trade. Whereas approximately 40 per cent of annual trade imbalances between any two CMEA countries was offset by a compensating imbalance in a subsequent year, the corresponding figure for the EEC (excluding Italy) was only 3 per cent.

The failure to achieve multilateral balancing emanates as much from the problems of the centrally-planned economies themselves, and the nature of the foreign trade system as from inadequacies in the CMEA monetary system.

The production conditions described in Chapter 3 have resulted in the CMEA becoming, according to Holzman (1979) 'a high cost, low variety, low quality production region'. Domestic price systems linked to average production costs tend therefore to overvalue CMEA products relative to those of the industrial market economies.

Under pure market conditions this would result in the CMEA countries tending towards balance of payments deficits with the West, which would be eliminated by a collective devaluation of the East European currencies relative to hard currencies, while they remained in approximate correspondence to one another. Quality differences between CMEA and western producers would be offset by price differences and the burden of adjustment would be borne by a reduction in real wages.

Holzman (1979) considers it unlikely that price reductions could stimulate sufficient sales to compensate for the lower quality of East European products, and it is highly probable that in practice any such adjustment process would be met by western nations by allegations of dumping. Consequently, the CMEA nations attempt to secure balance in trade with the West by the use of physical barriers, exercised by the foreign trade monopoly. The official exchange rates of the East European currencies remain overvalued in terms of hard currencies (which accounts for the existence of special exchange rates for tourists, visitors, and so on, which tend to approximate a genuine market rate) and the prices of internationally traded commodities are equated with domestic prices by taxing imports and subsidising exports.

In addition, the financial inconvertibility of the transferable rouble means that there is no direct link between exchange in intra-CMEA markets and world markets. A surplus acquired in intra-CMEA trade cannot be used to acquire products outside the bloc, nor need a deficit be paid for in terms of hard currencies. The CMEA region is therefore, as Holzman (1979) argues, 'a trade diverting customs union'.

The absence of commodity convertibility also means that surpluses acquired in intra-bloc trade cannot be used to acquire any commodity which the surplus country may wish to acquire and therefore the provision of exports cannot of itself guarantee a claim on the resources of a third country either inside or outside the bloc. The transferable rouble is effectively a unit of account, the international equivalent of passive money. The operation of passive money in the domestic economy requires the existence of a superior planning authority which can instruct (compel) one enterprise to deliver commodities to another enterprise. While no such supranational authority exists for intra-CMEA trade, countries operating on the principle of national self-interest will only enter into trade agreements if the value of commodities received in exchange exceeds the opportunity cost of those it provides.

The willingness to participate in multilateral clearing arrangements will therefore be critically affected by the degree to which prices used in the clearing arrangements reflect real opportunity costs. A nation's self-interest will be best served by exporting to other CMEA nations those commodities for which the opportunity cost (measured either by the domestic marginal cost of production or, where supply is inelastic, the price it could obtain for the commodity on world markets) is lower than the intra-CMEA price and by importing those commodities for which the opportunity cost (similarly measured) is higher than the intra-CMEA price.

Pricing in Intra-CMEA Trade

The most comprehensive western study of the process of price formation in the CMEA has been conducted by Hewett (1974), who provides some clear points concerning the *principles* of price determination which may be summarised as follows:

(1) The rules of price determination are established multilaterally, but actual prices are negotiated bilaterally.
(2) Some variant of world market prices (wmp's) is used.
(3) Prices are intended to be fixed or stable over a predetermined period.

In 1949 it was established that prices should be based on current world market prices and be fixed for the period of the annual trade protocol. The outbreak of the Korean War saw a rapid increase in wmp's, particularly for raw materials, and intra-CMEA prices were frozen at the 1949-50 level until 1956.

An attempt to establish rules by which CMEA prices should be governed was initiated in 1958 with the publication of the CMEA price clause, which Hewett argues delineates the areas within which negotiators may manoeuvre. The basic principles of price formation are that prices should be based on some variant of a previous period's wmp's 'cleansed of the influence of cyclical, speculative, monopolistic and other factors of a non-productive nature' (Ribalkin, 1978, p. 112) and that a single price should prevail throughout the bloc for a single commodity. In principle, prices should be sufficiently stable to eliminate the influence of short-term and cyclical fluctuations in wmp's, but sufficiently flexible to accommodate underlying changes in world supply and demand conditions.

This has led to debates concerning the period over which fluctuations should be considered cyclical or permanent. In practice, prices were supposed to be linked to a 1957 world price base from 1958 to 1965 but for 1966–70 were linked to an average world price for 1960–4, while a similar formula based on average prices, corrected by bilateral negotiations, prevailed from 1971 to 1974 (Ribalkin, 1978, p. 112).

Following the increase in world energy prices it was agreed at a standing commission of the CMEA that certain prices for 1975 should be based on an average of wmp's for the period 1972–4, and that thereafter prices should be changed annually on the basis of the preceding five years' wmp's. This formula has been renewed until 1985 (Ribalkin, 1978, pp. 112–26). Finally, prices obtained in this fashion are converted into transferable roubles at the official exchange rate.

In practice, the process is far more complicated. Trade negotiators come armed with documentation to establish what the appropriate world market price for a commodity is, which involves problems of identifying which market is to be considered the 'world' market, the time at which the 'world price' is fixed and the most appropriate product for comparison (Ribalkin, 1978, pp. 116–20). Effectively, prices result from a process of bilateral negotiation between foreign trade ministries who may be willing to cede a price disadvantage on one commodity in order to gain an advantage elsewhere (Ausch, 1972, p. 81). In practice, negotiators will be concerned not so much with the prices of individual commodities but with the 'total package' of the trade deal.

Ausch and Bartha (1968) undertook an empirical investigation of actual price determination in the CMEA in the mid-1960s and found a considerable degree of price variation for identical commodities traded between Hungary and other socialist nations, as well as with non-socialist nations. They found that this variation was far greater than that to be found from similar tests performed with French trade statistics or Hungarian pre-war trade statistics. Furthermore, in 1964 intra-CMEA prices diverged considerably from the wmp's on which they were supposed to be based, to the extent that machinery and equipment prices were 25.9 per cent greater than wmp's, raw materials 15.4 per cent greater and agricultural goods 1.7 per cent greater. They concluded that the divergence of machinery and equipment prices was largely explained by the fact that it is far more difficult to establish documentation for wmp's for finished goods than for raw materials, and consequently actual intra-CMEA prices for machinery and equipment were not wmp's at all, but were based on some assessment of past domestic production costs.

The divergence between prevailing intra-CMEA prices and world market prices was therefore greater for manufactured commodities than for basic commodities such as raw materials and agricultural produce, and this divergence was the opposite of that which would have resulted from supply and demand pressures within the CMEA and operated to the detriment of the less-developed. agrarian countries and suppliers of raw materials (principally, Romania, Bulgaria and the USSR).

Furthermore, the prices that CMEA countries obtain for manufactured commodities are frequently below those pertaining in trade between market economies (OECD, 1980).

Within the CMEA the policy of priority for heavy industry and the concentration on the production of steel and engineering products has caused excess supply of machinery and equipment, particularly of low quality, causing one CMEA economist to observe that 'there will be continuing shortages of modern, quality products and a surplus of obsolete, low quality products on the socialist market for machinery' (Pecsi, 1981, p. 91). Simultaneously, demand has increased for raw materials and energy products while the neglect of agriculture has also resulted in shortages of foodstuffs.

CMEA economists, therefore, distinguish between hard goods (those which are relatively underpriced on CMEA markets and can be sold on world markets for hard currency) and soft goods (those which are overpriced and difficult to sell on world markets). The degree of hardness differs from product to product and will change from time to time as supply conditions alter. The hardness of foodstuffs will be affected by harvest conditions which may cause countries to use hard currency reserves to import from the West, while improved quality standards for engineering products and even consumer goods will result in an increased degree of hardness of the commodities concerned. At the beginning of the 1980s, products were considered to range in the degree of hardness from raw materials and fuels, modern high-quality equipment and spare parts, engineering products that the importing country cannot manufacture itself, foodstuffs, products of light industry down to just about all other commodities (Pecsi, 1981, p. 137). The divergence of intra-CMEA prices from the opportunity costs of commodities traded within the CMEA is a principal reason for the failure of multilateral clearing systems and for the related phenomenon of the bilateral balancing of trade flows of specific commodities. Exporters will be reluctant to deliver hard goods into a multilateral accounting system where they may be offset by imports of an equal notional value but a far lower real value and will prefer instead to export such items to market economies where they will be paid in hard currencies which can be used to buy other hard goods. For the same reason importers will be reluctant to receive soft goods.

Trade negotiators confront each other in a position in which they will only be willing to export hard goods if they can guarantee receiving hard goods in exchange; negotiators cannot concern themselves with matters relating to the supply of exports and imports in isolation, but must ensure that the supply of exports will result in an inflow of commodities of greater or equal value to the commodities exported. Practically, this can only be achieved by bilateral negotiations where details concerning the supply of exports can be directly related to details concerning the supply of imports.

In practice, a multi-tier negotiating system has emerged in which, not only are hard goods largely traded in exchange for other hard goods, but certain categories of goods are only exchanged for similar items. This is largely a result of the domestic price system, which not only leads to differing degrees of hardness but makes it difficult to identify real production costs and compare the value of heterogeneous products. The greater the difficulty in identifying real production costs the greater will be the tendency to balance trade bilaterally in narrowly-defined product groups according to administrative production categories.

Settlements of intra-CMEA trade in hard currencies have been increasingly used in the 1970s to bypass some of the defects of the CMEA's monetary organisation. Such settlements accounted for 8 per cent of Hungary's trade with socialist countries in 1977–8 (Pecsi, 1981, p. 127). Hard currency settlements largely apply to trade in hard goods that cannot be balanced bilaterally, and frequently are additional to deliveries agreed in annual and five-year protocols and are conducted at current wmp's. Such settlements introduce an element of convertibility and multilateralism into intra-CMEA trade, but fail to tackle the underlying problems of pricing and the production of soft goods and by offering a pragmatic solution to a specific problem may delay a more thorough reform of the system.

Does the USSR 'exploit' Eastern Europe? Problems of Pricing in Intra-CMEA Trade

Considerable attention in the West has concentrated on the question of whether the USSR 'exploits' its East European satellites. A harder question to quantify is whether or not the East European economies benefit from their participation in the Soviet trading systems *per se.* To answer this question would involve the construction of an elaborate economic model that would have to go beyond the distributional questions of intra-CMEA pricing and credit relations, and would have to measure the effects of such factors as trade creation within the CMEA, trade diversion both from the West and from other East European countries to the USSR (the radial pattern), the potential lost economies of scale resulting from isolation from western markets and from western technology, the static losses but possible dynamic gains resulting from bilateral balancing and bilateral commodity balancing, the levels of international capital and labour flows, and so on.

The construction of such a model would also require making critical assumptions about a number of political as well as economic questions, including how magnanimous the Allied governments could have been towards the former Axis powers and in particular how they would have regarded the emergence of a strong central European economic union; the preferences the Balkan nations could have expected to receive for the sale of their agrarian products (particularly by the EEC); the real possibilities which may have existed for the expansion of trade between Eastern and Western Europe.

The size of the problem and the limitations of the available data have forced western researchers to concentrate most of their attention on improving knowledge of the operation of specific aspects of CMEA integration. Using partial data to assess larger questions of the costs and benefits of participation in the CMEA is hazardous, particularly in view of the fact that in any form of negotiations, negotiators may cede certain issues in order to obtain a concession elsewhere. Furthermore, such concessions need not be related to visible (or recorded) trade flows.

A result of the absence of an active intra-CMEA monetary unit and the inherently non-mercantilist, non-Keynesian, autarkic nature of CPE's is that any form of exports to another CMEA country, particularly of hard goods, tends to appear to the population as a whole as appropriation (Wiles, 1968, p. 451). Consequently, there is a wealth of anecdotal evidence to indicate that one country (normally the USSR) exploits another by 'taking' for example its food or natural resources, or does not pay a just price for the imports it receives.

A critical, and necessarily subjective question, when evaluating the fairness of intra-CMEA trade, is the matter of the values (eg, domestic production costs, socially necessary labour, world-market prices) that should be used to measure trade flows.

As intra-CMEA prices are theoretically based on world market prices (cleansed of monopoly factors) and may be considered to represent the opportunity cost (in terms of lost hard currency earnings) of exports within the bloc (or hard currency savings) from intra-bloc imports, many western analysts have accepted world market prices as a proxy for the fairness of prices in intra-CMEA trade (particularly for hard goods). An alternative is to estimate the prices CMEA countries actually obtain for their exports to the West or pay for their imports.

Even if this subjective standard is accepted, comparisons with world market prices, or prices actually prevailing in East-West trade, can be hazardous, and pose considerable problems of interpretation.

As a result of bilateral negotiations the prices of individual items cannot be taken as an indication of the fairness of trade in total. This could result purely from the overvaluation of the transferable rouble or from negotiating effects. If, for example, Bulgaria consistently pays 30 per cent more than world market prices for its imports, but simultaneously receives 30 per cent more than world market prices for its exports, then no exploitation can be considered to have taken place. (This conclusion only holds, however, if the transferable rouble is not financially convertible, or if as under existing IBEC rules each country's trade must be balanced within the region as a whole, otherwise surpluses obtained within the bloc could be used in world trade to obtain a greater volume of imports.) Consequently it is the net barter terms of trade — the terms at which exports in the aggregate are exchanged for imports in the aggregate — that must be used as the criterion of equity in trade.

Even here, comparisons with current wmp's can be misleading, and may reflect the impact of agreements made in a preceding period, or of bilateral agreements which may involve the exchange of products at different periods of time.

In the late 1950s Mendershausen (1959) showed that the prices that the East European countries obtained for their exports to the USSR were significantly lower than the prices the USSR paid for similar products imported from Western Europe; at the same time, the prices East European countries paid for imports from the USSR were higher than the price the USSR received for exporting similar commodities to Western Europe. Mendershausen concluded that the USSR, possibly unwittingly, used its position as dominant supplier and purchaser in the bloc to exercise price discrimination at the expense of the smaller CMEA members.[4] The counterargument developed by Holzman (1962a, 1962b, 1965) that prices prevailing in East-West trade may reflect western discrimination against East European countries and that similar calculations could demonstrate that the smaller countries were simultaneously exploiting the USSR can be illustrated by the hypothetical example set out in Table 8.3.

Table 8.3: Hypothetical Prices in East-West and Intra-CMEA Trade

	World market price ratio	Intra-CMEA price ratio	East-West trade prices	
			CMEA exports	CMEA imports
(A) Before 1975				
Unit of oil	30	35	20	40
Unit of manufactures	30	32	20	40
(B) After 1975				
Unit of oil	60	40	50	60
Unit of manufactures	35	35	20	40

Assume *before 1975*:

(i) The USSR exports oil and imports manufactures from both East Europe and 'the West'.

(ii) Eastern Europe exports manufactures to the USSR and to 'the West'.

(iii) Eastern Europe and the USSR both have to pay more for their imports from hard currency sources, and receive less for their exports to hard currency markets, than the prices that prevail in trade between market economies.

Under these circumstances:

(i) Eastern Europe will sell manufactures to the USSR at a price which is lower than the price the USSR pays for manufactures from the West (32 against 40).

(ii) Eastern Europe will buy oil from the USSR at a price that is higher than the USSR obtains for its exports to Eastern Europe (35 against 20).

This is the position observed by Mendershausen (1959) from which he concluded that the USSR discriminated against Eastern Europe.

But Holzman (1962a) observed that it will also happen that:

(iii) The USSR will sell oil to Eastern Europe at a price which is lower than Eastern Europe pays for oil from the West (35 against 40).

(iv) The USSR will buy manufactures from Eastern Europe at a price which is higher than Eastern Europe obtains for its exports to Western Europe (32 against 20).

Holzman argued that it would appear that Eastern Europe exploits the USSR, while the USSR simultaneously exploits Eastern Europe, which is clearly an impossibility.

Holzman (1965) demonstrated that this situation actually existed in the late 1950s by showing that the prices that the USSR and East Europe received for exports to Belgium and the UK were lower than those prevailing in trade between the two countries, while the prices the USSR and East Europe paid for imports from those countries were actually higher than those prevailing in trade between Belgium and the UK. He concluded that part of the explanation for the situation described above was that the West discriminated against the CMEA nations.

Holzman (1962a) also proposed that prices and costs in CMEA diverged from those on world markets due to the 'customs union' effect. Because the CMEA is a trade preference area, with a penchant to autarky, and remains isolated from many capitalist markets (Stalin's two world markets) relative costs will differ from those prevailing in capitalist markets. The conclusions that arise from this analysis are of far more than academic importance, and have been the basis of serious disputes between Soviet and East European economists and even lie (indirectly) behind Czechoslovakia's actions in the mid-1960s, and are of considerable significance for western trade policy towards Eastern Europe.

The argument can be illustrated from Table 8.3 by taking the not unrealistic assumption that the real costs of production in the CMEA in terms of factor inputs, are higher than those prevailing in western market economies, and that while appropriate amounts of manufactures and oil cost 30 units to produce in the West the equivalent amounts actually cost (in terms of factor inputs) 32 and 35 units respectively to produce in the CMEA. If Eastern Europe produces manufactures and exports them to the USSR in exchange for oil at world market prices (30:30) the USSR will effectively lose in its trade with Eastern Europe, particularly if it were capable of producing manufactures at the same real cost as Eastern Europe. MacMillan (1973) argued that this situation had actually occurred in the late 1950s, and that the USSR actually made real losses in its trade with Eastern Europe. Using the Soviet 1959 input-output table he estimated that the domestic resource cost of producing items for export to Eastern Europe actually exceeded the production costs the USSR would have incurred if it had produced domestically the commodities it imported from Eastern Europe.

If, however, intra-CMEA trade were to be conducted on the basis of real production costs within the CMEA ('own price base') rather than world market prices, then it can also be seen from Table 8.3 that:

(v) The prices (net barter terms of trade) of manufactures for oil (32:35) would be less favourable to Eastern Europe than those based on world market prices (30:30).

(vi) The East European countries would still receive better terms of trade for manufactures against oil within CMEA (32:35) than those prevailing in their trade with the West (20:40).

On the basis of this example, it would be more beneficial for the East European countries to trade within the CMEA than with the West when all factors are taken into account, an analysis which would be consistent with the 'customs union effect'. After 1975, following the increase in world oil prices, the introduction of the sliding world average price system into intra-CMEA trade may be more similar to that shown in Section B of Table 8.3 (see Tardos, 1978). The intra-CMEA price for oil and manufactures has increased, but the increase for oil is far greater than for manufactures; Eastern Europe's export prices for manufactures have actually worsened relative to world market prices because of the low quality of East European exports (Tardos, 1978, pp. 264–5), but the price ratio of oil sold in the West to manufactured imports from the West has improved.

As a result, the relative prices pertaining for East European manufactures in exchange for Soviet oil (35:40) are now far more favourable than those pertaining in East-West trade (20:60) while the USSR receives far more favourable terms of trade with the West (50:35) than in intra-CMEA trade (40:35).

The critical question of *why* the CMEA is a high-cost production region and why it obtains such unfavourable prices in East-West trade would still remain.

If the answers to these questions lie in the nature of the economic systems and the structure of organisation of the CMEA itself, then these factors should be considered an economic cost to the East European countries arising from their membership of the Soviet bloc. If on the other hand these factors result from western economic discrimination, then participation in the CMEA offers benefits to the East European countries that are unobtainable elsewhere.

In either case, the pursuit of policies in the West that increase the price disadvantage of trading with the West to Eastern Europe will reinforce the 'customs union effect' and increase the potential benefits of trade with the USSR to the East European countries.

Notes

1. See Baykov (1946), p. 60.
2. For a discussion of the distinction between financial and commodity convertibility see Wiles (1973).
3. For a distinction between 'random reciprocity' and planned bilateralism see Wiles (1968), pp. 254–6.
4. For a detailed analysis of the debates on the 'fairness' of intra-CMEA trade, see Wiles (1968), Chapter 9 and Hewett (1974), pp. 43–57.

9 ECONOMIC INTEGRATION AND THE COUNCIL FOR MUTUAL ECONOMIC ASSISTANCE (CMEA)

The Origins of the CMEA

Although official accounts report that the CMEA was established in 1949 following a conference in Moscow from 5 to 8 January attended by representatives of Bulgaria, Hungary, Poland, Romania, the USSR and Czechoslovakia, the exact date of foundation and the principles on which agreement was reached is still the subject of debate.[1] In the years between its foundation and the death of Stalin, the CMEA's activities were restricted to developing unified systems of statistical reporting, collating members' plans and recording trade. There was little or no attempt to integrate the economies of the member countries who developed their own parallel industrial structures and systems of organisation and drew up their domestic economic plans with little mutual consultation. As foreign trade negotiations were not concluded until production plans had been finalised, trade flows were largely confined to the exchange of products that were surplus to domestic requirements.

There is considerable speculation in the West on Stalin's reasons for establishing the CMEA if he originally intended that the institution should become moribund. Political scientists propose that the foundation of a socialist international economic authority was required for propaganda purposes following the Soviet veto of East European participation in the Marshall Plan and as a means of excluding Yugoslavia from international socialist co-operation following the Tito-Stalin rift. Kaser (1967, pp. 21–6) argues that the quiescence of CMEA coincided with the demise of Vosnesensky, who, as Chairman of the USSR Gosplan from 1937 to March 1949, had latterly been attempting to bring about domestic economic reform which would have placed greater emphasis on market signals and active money. The logical inference is therefore that the CMEA would also have allowed market signals into international relations between the communist states along the lines contained in the proposals to establish customs unions (and which would have considerably simplified East European contacts with Yugoslavia). Wiles (1968, p. 313) agrees with this explanation and concludes that following the denunciation of Vosnesensky and the reversal of his policies 'Stalin found himself saddled with an organ he had allowed to be born, but could not personally work with.'

Van Brabant (1980, p. 51) argues that the political and economic views of the senior participants in the January 1949 conference (especially Suslov, Hilary Minc, Erno Gero) make it unlikely that they would have agreed to proposals for integration on market lines. It is also interesting that Kosygin, who subsequently became the major architect of proposals for improved integration on the basis of plan co-ordination rather than market levers, was considered by Kaser (1967, p. 23) to be Vosnesensky's major ally in promoting economic reform in the USSR, which caused Kosygin to lose his post as Minister of Finance in January 1949. Van Brabant (1980, p. 41) cites the Hungarian economist Friss in support of the hypothesis that, from its foundations, the main method for economic integration was considered to be plan co-ordination, and that the Commission was actively involved in preparing proposals to implement plan co-ordination (together with proposals for multilateral clearing, price formation and scientific co-operation) at the beginning of 1950. Van Brabant proposes that Soviet support for the work of the CMEA ceased about the middle of 1950, at about the time of the Korean War as a result of which the CMEA became inactive while the individual countries were required to pursue their separate industrialisation programmes and contribute to the war effort.

As a result, therefore, each East European country developed its own heavy industrial base, duplicating production in other East European countries and simultaneously developed separate economic structures with vertical decision-making processes. Economic relations between the countries were not based on direct contacts between enterprises, but resulted from the flow of information and instructions up and down the respective national economic hierarchies which were only linked with one another at the centre by the individual ministries of foreign trade. Individual national plans were largely constructed in ignorance of the production and supply possibilities existing in other CMEA countries, while the absence of currency and commodity convertibility removed any incentive for one country to expand its domestic production base in order to supply other countries. During Stalin's lifetime the costs of this system were almost entirely borne by the East European economies which were hardly integrated with one another and were dependent on the Soviet Union for supplies of raw materials, while the former Axis powers also had to pay severe reparations for war damage. Following Stalin's death and the end of the Korean War this industrial and organisational parallelism posed severe difficulties both for the individual East European economies and for their economic integration. Although the Soviet leadership, and in particular Khrushchev, and later Kosygin, attempted to stimulate a greater degree of economic integration, they were unwilling to move far towards the introduction of market levers, and clear differences of opinion arose between the members of the CMEA on the most appropriate method for integration.

Institutional Factors Affecting Economic Integration in the CMEA

1. The active members of the CMEA consist of a dominant nation, the USSR with a population of 270 million; six European nations, Bulgaria (10.8 million), Hungary (10.7 million), GDR (16.7 million), Poland (35.2 million), Romania (22.0 million), and Czechoslovakia (15.2 million); three developing nations, Vietnam (52.4 million), Cuba (9.7 million) and Mongolia (1.6 million). It is not therefore a group of developed industrial economies, but is far more heterogeneous in terms of industrial structure, income levels and cultures, whilst the dominant nation itself is not the richest in terms of income per head in the bloc and contains considerable diversity within its boundaries.

2. The original members largely constituted those nations that had either supported the Axis powers in the Second World War or had been liberated from the Axis powers with the assistance of Soviet troops. Albania, who joined the CMEA in February 1949 and has ceased paying contributions and participating in its work since the end of 1961, appears to be an 'inactive member' without any remaining rights or obligations.

3. The CMEA nations are basically self-sufficient in most raw materials and sources of energy. These resources are unevenly distributed throughout the bloc and the greatest resource potential exists in West Siberia, whose location and harsh climate result in high extraction and transportation costs.

4. The member countries have all adopted the Soviet system of central planning and in addition to coping with pragmatic questions of economic development are considered to be in

transition to a higher stage of economic development — full communism.

These factors combine to provide a contrast to traditional theories of empire in which technologically advanced economies have sought markets and sources of raw materials through trade with, and investment in, countries at a lower stage of economic development. The politically and geographically dominant nation in the CMEA, the USSR, is rich in natural resources but technologically less developed outside the space and defence sectors than several of the East European nations that became its satellites.

National Attitudes to Integration in the CMEA: Plan and Market Approaches

Debates between Soviet and East European officials and economists specialising in CMEA affairs have largely concentrated on such questions as the methods and terms of provision of Soviet supplies of energy and raw materials and on the methods of organisation and integration to be used in the CMEA, on the increased use of plan or market levers and on the related question of whether such methods help or hinder trade outside the bloc.

Frequently the arguments proposed represent the national interests of the individual countries and indicate that careful consideration has been given to the material impact of any given proposal on the country concerned. Integration proposals also tend to reflect the nature of organisation of the domestic economy, with the most consistent proposals for the increased use of market levers being proposed by Hungarian economists.

Some cautionary points should be noted:

1. In view of the USSR's dominant political and economic position and its role as the major supplier of raw materials to the East European countries, it is tempting to seek a straightforward dichotomy of interests between the USSR on the one hand, and the East European states on the other. Although the USSR has made the most consistent proposals for integration by the use of methods of joint planning, there is little evidence of a unity of views amongst the other East European states. In some circumstances, differences of opinion *between* the East European states (particularly between the more industrialised Central East European and agrarian South East European states) have been bitter, and the

USSR has been required to act as an arbitrator as much as a defender of its national interest. Under these circumstances informal alliances may be formed between member states that will not hold over to other issues.

2. State officials and negotiators who represent national interests derive their authority from the continued existence of the nation state. Consequently, those responsible for formulating policy retain an interest in the preservation of domestic government authority and the continued existence of the nation state. This factor is complicated in the CMEA by the existence of state authorities in the economic sphere — national planning authorities, national material-technical supply agencies, ministries of foreign trade, and so on, while those responsible for proposing and implementing policies in the CMEA represent those national organisations. There may come a point at which the pursuit of integrationist policies (in the form of the increased use of market levers or proposals for supranational agencies) conflict with the preservation of the power base of those responsible for their implementation. There is a powerful impetus to preserve the status quo, even where the status quo conflicts with the economic interests of the nation state.

3. On the other hand, national consciousness also extends to the fear of incorporation. In particular, the small industrialised East European nations that may have the most to gain in economic terms from improved integration (in particular the benefits arising from economies of scale) may feel themselves to be vulnerable in terms of the loss of nationhood, particularly where integration proposals imply closer ties with the USSR which has incorporated formerly independent nation states into its borders and its planning system. One solution to this problem is to seek greater integration between the East European states and to pursue an increased role for market levers in the process of integration. The problem of supplies of energy and raw materials (to be purchased either from the USSR or outside the bloc) remains, however, and is a powerful source of fragmentation.

These problems have resulted in different attitudes towards integration between the East European countries and attitudes within individual countries have been affected by changes in political circumstances.

As a very rough generalisation, Hungarian, Czechoslovak (particularly in the mid-1960s) and Polish proposals have placed greater emphasis on the use of market levers as a means of integration, on greater integration between the East European states, and on a greater role for trade outside the bloc. At the other extreme, Romania has placed greatest emphasis on the desire to preserve the power of domestic central-planning authorities, while simultaneously rejecting attempts to introduce any form of supranational authority and integration through market levers. As a result, Romania has effectively been forced to seek greater trade links with the West. Finally, Bulgaria, and to a lesser extent the GDR, have placed more emphasis on plan levers and have been more receptive (especially since 1968) to Soviet proposals for integration through the use of plan levers.

Alternative Methods of Integration. Plan and Market Solutions

Two theoretical extremes can be proposed for stimulating integration within the CMEA — the full market solution or the full centrally-planned solution. The full market solution would decentralise production decisions to enterprises who would draw up production plans on the basis of profit criteria. Prices of outputs and factors of production would be determined by supply and demand, commodities and factors of production (capital and labour) would be free to flow across national boundaries and enterprises would be permitted to choose to purchase inputs and sell outputs either within the domestic economy or from/to other bloc countries. If external optimality were to be introduced enterprises would have to adjust their input and output decisions according to external opportunity costs, and as a result intra-bloc prices would have to reflect world market conditions.

A critical problem would still concern the role of national and supranational planning agencies. Central authorities would still be required to stimulate investment, maintain full employment (or minimise the effects of frictional unemployment resulting from sudden supply and demand changes), ensure reasonable income equality, control the money supply and maintain external equilibrium. Full market logic would appear to indicate that these policies should be conducted by supranational authorities responsible for the CMEA as a whole on the basis of a single currency. Under these circumstances the pursuit of income equality would require the redistribution of income from the richer citizens in the bloc across national boundaries to the less wealthy.

Alternatively, the whole paraphernalia of separate national currencies, national central banks and national planning agencies could be retained with trade between member countries taking place on the basis of full currency and commodity convertibility and exchange rates determined by market factors. The major role for CMEA agencies would be to ensure policy harmonisation.

The full centrally-planned solution would involve the establishment of a single supranational central-planning agency capable of drawing up and implementing plans on a bloc basis, a single material-technical supply agency covering deliveries within the whole bloc and a single ministry of foreign trade responsible for all extra-bloc trade activities. Money would remain passive in inter-enterprise transfers (including international transfers) and a single monetary unit would facilitate enterprise accounting, although separate currencies (with fixed exchange rates) could be retained in the consumer sector as a gesture towards national identity.

Soviet and East European economists have not proposed either the full market or full centrally-planned variant as practical solutions to the problems of integration in the CMEA, but have attempted to adjust the existing system by the increased use of market levers or plan forces. In particular, most proposals have involved the retention of separate national planning and banking authorities.

Consequently, proposals for improved integration based on plan instructions, undertaken on the basis of separate centrally-planned economies, require that economic decisions must be examined on a bloc basis before members' domestic plans are formulated. Ultimately, the degree of integration reached will depend on the time horizon over which production decisions are affected by bloc rather than purely national considerations, ranging in the short term from influencing current operational plans by attempting to affect the level of output from existing plants to, in the longer term, introducing investment proposals based on specialisation agreements into members' national plans and drawing up comprehensive investment proposals based on joint participation and construction. Such proposals are frequently referred to as 'production integration', as members agree to specialise in the production of certain items and draw up their domestic production and foreign trade plans accordingly. The major purpose of such integration is to avoid duplication in investment production, and research and development, and bring about economies of scale.

There are areas where the use of market levers and production integration may be considered complementary to one another and others where they come into direct conflict.

The critical area of conflict between market and plan levers arises out of proposals that involve full commodity convertibility for national currencies. If an enterprise in one country is free to bypass both its own central authorities and the central authorities in other CMEA countries, the central-planning authorities and material-technical supply agencies no longer retain complete control over the flow of resources within their respective economies. Full commodity convertibility is therefore incompatible with detailed operational planning conducted on the basis of national material balances, and market levers must play a far greater role both in the process of integration and in determining members' production plans, while central planning is broadly restricted to macro-economic decisions.

Provided market levers remain subordinate to plan directives, and do not involve commodity convertibility, they may be regarded as an aid to efficient plan co-ordination by permitting rational calculation on which specialisation agreements may be based. To fulfil this role, prices should reflect the real costs involved in specialisation and investment decisions. In joint investment projects, for example, rational prices would act as a guide to the calculation of all costs and benefits involved in a project, the contributions of participating countries and their payment in the form of products resulting from the investment, while all production decisions would actually be taken by the planning authorities of the individual nations. Passive money balances (in transferable roubles) would also act as a store of value when nations' contributions to and receipts from any agreement are not in balance in any accounting period.

Under these circumstances, market and plan levers are complementary and as a result there is a degree of shadow-boxing in debates on the nature of integration in the CMEA which leads to considerable ambiguity in the agreements that are reached. All parties can safely agree that the increased use of money levers and plan co-ordination are essential to the process of integration, while reserving their position on which should predominate.

Soviet Political Proposals and Incorporation

The incorporation of socialist states into an economic union regulated by a joint plan has been a consistent theme in Soviet theoretical and political writing. The programme of the Communist

Party of Russia, adopted at the Eighth Party Congress (March 1919) called for 'the maximum union of the whole economic activity of the country, in accordance with one general state plan'; article 96 of the programme also called for a single unified plan to co-ordinate activity with other socialist countries: 'We must promote a close economic collaboration and a political alliance with other peoples, simultaneously striving to establish a unified economic plan in conjunction with those among them that have already established a Soviet system.'[2]

Although this could be interpreted as referring to relations with those republics that ultimately became part of the Soviet Union, the proposal was clearly extended to 'any country in which the proletariat gains the upper hand' by Bukharin and Preobrazhensky in the *ABC of Communism* (first published in 1921). They proposed that integration should extend beyond pure trade relations (trade integration) to factor integration (via factor mobility) and policy integration under the direction of a common economic plan:

> We must aim not merely at economic exchanges with such countries, but if possible we must collaborate with them in accordance with a common economic plan. Should the proletariat prove victorious in Germany we should establish a joint organ which would direct the common economic policy of the two soviet republics. It would decide what quantity of products German proletarian industry should send to Soviet Russia — how many skilled workers should migrate from Germany (to the Russian locomotive factories for instance) and what quantity of raw materials should be sent from Russia to Germany. (Bukharin and Preobrazhensky, 1970, p. 326.)

Referring to the question of political incorporation in 1972 Brezhnev proposed:

> When the question of uniting the Soviet Republics in a single union of SSR's arose 50 years ago, Lenin pointed out that the Union was necessary . . . to accomplish the peaceful creative tasks of socialist construction more successfully by common effort. *In principle, the same applies to the fraternal community of Socialist states that belong to the Warsaw Treaty and the CMEA*. (Rakowska-Harmstone, 1976, p. 42.)

Stalin, however, had assiduously avoided such an outright annexation, but had also vetoed any separate customs union involving groupings of East European states; only Khrushchev (1962b, p. 111) in a speech to the Central Committee Plenum of the CPSU on 19 November 1962 was to go so far as to propose a supranational planning organ when he declared:

> It is necessary for us to go quickly to the establishment of a common single planning organ for all countries, which would be composed of representatives of all countries coming to CMEA. This planning organ must be composed of such people that are fully empowered to put together common plans and resolve organisational questions in order to co-ordinate the development of the economies of the countries of the socialist system.

Integration measures proposed by Kosygin have carefully avoided any implications of supranationality. Soviet attitudes towards the longer-term political question of the integration of the CMEA have been summarised by three Soviet CMEA specialists (Alampiev, Bogomolov and Shiryaev, 1973, pp. 82–5), who propose that the integration process will take place in three stages: In the first stage,

> the socialist countries' national economies and state property will remain the main arena of independent reproductive processes, even if there is to be a steady growth within these processes of the role of the world socialist economy, especially of its integrated part. There will be an extension of the sphere for deploying material, financial and manpower resources on the scale of the whole community . . . selection of this or that economic solution will be determined by considerations both within the given country and the framework of the community.

At the second stage of integration,

> from plan co-ordination the countries will go on to the formulation in one form or another of a common plan . . . this will require new forms for joint forecasting and planning to include the possible establishment of international planning bodies.

But, they warn,

> achievement of the final result — integration of all the socialist national economies in one international economy regulated under a common plan — goes beyond the framework of the foreseeable future, and it would be a gross error to regard the strategic goal as being the goal of the immediate measures designed no more than to create the pre-requisites for setting the integration process in motion.

Hungarian proponents of market type integration argue that the planned approach corresponded to the stage of extensive growth, in which bilateral negotiations predominated, and if supranational authority is to be avoided greater use must be made of money and price levers in the process of integration.

The Process of Plan Co-ordination and Disputes within the CMEA

The Question of Supranationality

It would be a mistake to view Khrushchev's proposals to bring about improved integration and co-operation in the CMEA purely as an attempt to impose Soviet views on a recalcitrant Eastern Europe — in practice, the USSR found itself in the centre of a dispute between the industrial and agrarian East European nations, in which it was ultimately forced to take sides, but when it did, attempted to implement policies of its own choosing.

The reversal of industrial priorities following the end of the Korean War, and the institution of the new course resulted in a reduction in demand for the engineering products of Czechoslovakia and the GDR and simultaneously for iron and steel, in particular. The less-developed economies, particularly Romania and Bulgaria, and to a lesser extent Poland and Hungary who had been required to develop their heavy industrial base, were reluctant to cut back on the output of their iron and steel industries. This placed considerable strain on bloc deposits of iron ore and coking coal which were primarily located in the USSR. During the mid-1950s attempts to introduce production integration were largely limited to affecting the levels of output from existing plants. Logically, structural imbalances in the economies could only be removed by influencing members' investment plans which, if not secured by voluntary policies or by market forces, required the use of supranational authority.

The problem ultimately centred on Romanian proposals to build a metallurgical complex situated at Galati on the Danube, close to the Soviet border. Romanian proposals were based on a form of economic nationalism which placed emphasis not just on industrialisation but on a 'many sided industrialisation' spread across a wide range of industrial products, involving the development of heavy industry with little regard to the domestic supply of raw materials. The theory of comparative costs has been attacked by Romanian economists (Puiu, 1969; Puiu and Ciulea, 1967) on the grounds that to pursue industrial specialisation on the grounds of *existing* comparative advantage would mean that Romania would remain a predominantly agricultural nation in perpetuity, dependent on the export of agricultural commodities and raw materials. On grounds that are similar to those advocated by Rauol Prebisch, they argued that in the long term the price of such exports would deteriorate relative to the price of imports of industrial commodities, leading to a slow-down in the rate of accumulation and growth. It was also argued that even though specialisation on the grounds of comparative costs could increase the productive capacity of the Romanian economy, it would simultaneously bring about greater benefits to the industrial economies and thus widen the absolute gap between them. (Similar arguments have been raised recently by Mongolian economists — Lufsandorj, 1978.)

Romanian economists argued that it was therefore necessary to raise the productive potential of the domestic economy (levelling-up) before specialisation could take place.

In 1958, Romania announced plans to substantially expand her iron and steel production (including the Galati complex) and more than double the output of machine tools (Montias, 1967, p. 54). Although plans were also announced to step up domestic prospecting for ferrous and non-ferrous ores (at the expense of oil prospecting), the siting of the complex made it apparent that the use of Soviet ores, as well as Soviet aid in the construction of the project, was intended. (These points have been confirmed by conversations with Romanian emigré economists.)

Although it is easy to cast Romania as the irrational or intransigent partner in this dispute, Montias (1967, p. 191) has demonstrated that she had been pushed into this position by the reluctance of the industrialised countries to forgo the production of technologically simple industrial processes such as tractors, lathes, ball bearings (in which the less-developed countries *could* have obtained a comparative advantage), while the more industrialised nations concentrated on the production of more complex industrial machinery.

Consequently, although the specialisation proposals implied benefits to the bloc as a whole, certain countries would clearly be disadvantaged by certain specific proposals and there was no guarantee that the benefits from any set of proposals would be equally distributed. The implementation of decisions based on bloc rather than national interests requires the development of a reward system that compensates losers for their acceptance of proposals that are not in their individual interests, or the ability to pursue proposals without the participation of disinterested nations, or the acquisition of supranational powers. The Hungarian economist, Ausch (1972, p. 71) has argued that the absence of commodity convertibility in the CMEA prevented the possibility of any feasible compensation process.

It appears that in the initial stages some East European leaders were willing to see the CMEA take on such supranational powers, although they may have baulked at Khrushchev's ultimate proposals. Kaser (1967, p. 108) provides an interesting quotation from Fierlinger, the President of the Czechoslovak National Assembly, who said in 1963:

> From the very start Czechoslovakia eagerly supported the CMEA in its efforts to promote economic co-operation. We welcomed Khrushchev's appeal for the closest economic unity of the socialist countries . . . that the CMEA play a bigger part met a warm response in Poland . . . We would like to have its recommendations and decisions (unanimously agreed on of course) binding on all member states . . . We need the enlistment of the most efficient and competent experts from all member countries . . . they would act independently of their respective national authorities and be responsive to the CMEA collective leadership.

The most serious attempt to cope with the problems was made in 1962 (Kaser argues that this resulted partly from the construction of the Berlin Wall, which limited the GDR's ability to export to the West). In June 1962 the Conference of Communist and Workers' Parties, including all the Party leaders approved 'The Basic Principles of the International Socialist Division of Labour'. The 'Basic Principles' contained no overt references to supranational planning, other than that specialisation would require the concentration of investment and production of certain commodities in one or several countries of the bloc, and more importantly that raw materials should be processed 'at the place that they are found' (Kohler, 1965, p. 384), which would effectively limit Romania's plans to develop an independent steel industry.

In June, however, the Soviet-Romanian Trade Communiqué referred to continued Soviet assistance in the construction of a metallurgical complex in Romania. Later, in an article in the *World Marxist Review* published in September 1962, Khrushchev (1962a) appeared to imply giving the CMEA supranational powers by arguing that it was necessary 'by starting out from planning at the national scale, to go in for planning at the level of the CMEA, and afterwards at the level of the socialist world system as a whole'. However, he diluted the proposal by suggesting that such planning would mainly consist of the composition of 'balances of production and consumption of the main types of manufacture . . . to arrive at a kind of overall balance that would fulfil the function of a joint plan' — a proposal which was entirely consistent with the 'Basic Principles'.

Khrushchev also proposed that monetary measures should be used to stimulate integration but that they should involve the use of 'passive' money responding to 'Agreed National Investment Plans which would take into account both national and common interests . . . and the joint-financing of industrial building, transport installations and other enterprises of international significance'. This could involve 'joint credits for enterprises which are being built or constructed by way of international specialisation and co-ordination. It may be expedient to establish a joint bank of the socialist countries for financing measures of this kind.'

On 19 November, Khrushchev (1962b, p. 111) proposed his 'common-planning organ'. The Romanian reaction was to convene a Special Party Plenum from 21 to 23 November, after which it was announced (25 November) that the construction of the Galati metallurgical combine would continue with the co-operation of an Anglo-French consortium (Fischer-Galati, 1967, pp. 90–1). Romania's dispute with the industrialised nations crystallised into a dispute with the USSR, which accelerated in the period up to Khrushchev's ousting and, contrary to expectations current at the time, has not abated since.

The Resumption of Market Reforms

The market-oriented reforms proposed in Hungary and Czechoslovakia were linked to proposals to place greater emphasis on importing, developing and diffusing western technology throughout the CMEA and exporting commodities to pay for the imports. This policy also had considerable implications for CMEA integration. On the one hand they could be seen as strengthening the case for production integration and a planned unified approach to western maufacturers, while on the other they also increased the need for a rational system for the comparison of cost structures, not just between different East European countries, but also between those countries and the West.

Hungarian economists in particular tended to view market-type reforms, technology imports from the West and improved CMEA integration through the use of market levers as complementary. CMEA specialisation based on imported technology would result in economies of scale that would simultaneously increase the attractiveness of the CMEA as a market for western producers, and would reduce the unit cost of licences and technology to eastern imports (Radice, 1981a, p. 121). Hungarian economists argued that optimal enterprise size would require that enterprises produce not for one market (domestic) nor two (domestic and socialist) but three (domestic, socialist and world) and would simultaneously require inputs from all three sectors.

Economic efficiency (and the use of profits as an enterprise success indicator) required that each enterprise should be able to compare accurately the costs of inputs purchased from each of these sources and the value of sales to each of these markets. If western suppliers were to react more quickly than their CMEA competitors, East European importers would tend to give them preference. In order to market output in the West it would be necessary to bring international quality specifications, technical levels and above all flexibility and speed of response to changing international market conditions to bear directly on enterprise decisions. Logically this required direct contacts between enterprises in the domestic economy (active money) and its counterpart in intra-CMEA trade — commodity convertibility. This would provide considerable problems for dealings between economies with centralised and market-type mechanisms. Either the centralised economy would have to accept decentralisation in its dealings with the decentralised, or the decentralised would have to accept centralisation in its dealings with the centralised.

Either circumstance could be damaging to the performance of one of the economies concerned. In addition, however, the increased role for active money and market forces and the decreased use of material balances in the domestic economy would increase the necessity for unplanned (spontaneous), foreign trade flows as a balancing factor.

Given the existence of a world (capitalist) market in which surplus and deficit raw materials can be bought and sold, this should present few problems. But if the socialist states felt it to be politically desirable to remain broadly autonomous in respect of raw material supplies it effectively meant that the USSR, as the principal supplier, would have to take on the role of the balancing factor in intra-CMEA trade, and have its domestic output plans influenced by decentralised decisions emanating elsewhere in Eastern Europe.

This must have posed something of a dilema for the Soviet leadership and its advisers. As the USSR was a high-cost producer (and transporter) of raw materials, existing world market prices provided little incentive to the USSR to invest in the development of its Siberian raw material base to meet the demands of the East European producers. Failure to meet those demands however would force those countries onto world markets to buy raw materials and energy supplies, which would in turn strengthen the demand for domestic reform to generate hard currency earnings and to bring world market prices and conditions directly into enterprise accounting and decision-making. At 1960 world market prices, therefore, the USSR had to contemplate a continued subsidy of raw material deliveries to Eastern Europe, sanction deep-seated reforms in those countries, or use diplomatic or other pressures to stop them.

Soviet attitudes towards market-oriented reforms and market-type integration in the CMEA were ambivalent in the period from 1964 to 1967. In November 1967 the Soviet chairman of the CMEA, Fadeev, wrote in *Pravda*: 'A number of socialist countries are carrying out economic reforms . . . it will be necessary to seek out new forms and methods of co-operation . . . through the wide use of economic levers', and in December 1967 the communiqué of the 21st CMEA Session referred favourably to the use of monetary levers as a method of integration.

A more muted reaction was also put forward in 1967 by the

leading Soviet theorist on CMEA affairs, Bogomolov (1967) who noted that several East European economists were proposing that central authorities should concern themselves with planning only the basic proportions of the national economy, leaving the estimation of detailed output nomenclatures to enterprises themselves, and that several economists were proposing that production specialisation decisions should be decided at a similar level. This would clearly have implications for those economies retaining a centralised approach:

> considerable differences exist in the forms of management between CMEA countries . . . the powers accorded to economic organisations and their economic independence vary from one country to another . . . this is important when talking of enterprises and associations being given the right to freely enter foreign markets and draw up direct links . . . eg in the GDR economically accountable trusts have already been formed . . . in the USSR it is still being decided . . . the new approach to implementing international production co-operation cannot be accepted without reservations.

It was important, he added, to ensure that direct links between enterprises were not to be left entirely to the spontaneous play of market forces, but were to be controlled either by the planning authorities of the countries concerned, or by CMEA organisations established for the purpose, such as Intermetal: 'This does NOT mean that direct links between enterprises, trusts and ministries of various socialist countries are to be left to themselves. They require definite forms of organisation and regulation.'

At the same time the attention of many Soviet specialists on CMEA affairs shifted towards the problem of pricing in the CMEA and, in particular, to the lack of incentive provided to raw material producers by the system based on world market prices, which effectively meant that the USSR was not compensated for the higher average and marginal costs involved in developing Siberian deposits, which were frequently exchanged at price ratios prevailing on world markets for machinery and equipment which were below international quality specifications.

Ladygin and Shiryaev (1966), for example, proposed the use of a price system based on intra-bloc cost ratios (own price base) rather than world market prices, a concept which was supported by

Bulgaria, and, it was hoped, would win support from Romania.

> Exporters of raw materials and food products often find it more profitable to produce manufactured goods domestically, even under less favourable conditions than those pertaining to the main exporters of those goods, than to import manufactures for raw materials at relatively unfavourable prices . . . it seems expedient to shift to a system . . . based on actual production costs in the CMEA countries.

Ladygin and Shiryaev also opposed the use of market forces, but attempted to assuage Romanian feelings by dismissing 'the spontaneous interplay of market forces' which together 'with efforts to create supranational directing agencies are equally unacceptable'.

This view was rejected by Czechoslovak and Hungarian economists, who argued that if implemented, the industrialised nations should obtain their raw materials from outside the bloc and redirect their exports accordingly.

Ausch and Bartha (1968) recognised the mathematical accuracy of Soviet arguments but rejected them on the grounds that they would ossify existing inefficiencies in intra-CMEA trade, and would not in practice create new resources but would be purely redistributive: 'At the same time we hear unrealistic demands that we change the price ratios of finished goods and raw materials and use the resources released to finance the production of raw materials.'

Hungarian economists continued to stress that reforms of the domestic economic systems required new methods of economic integration. The polarisation of CMEA economists on national lines was clearly reflected in the papers presented to a CMEA conference called to discuss methods of integration in 1970. Hungarian and Polish economists in particular stressed the need for 'non quota, free turnover' based on direct relations between enterprises, and not subject to central control.

These proposals were strenuously resisted by the Soviet economist Bogomolov (1973, pp. 31–6) who stressed the

> concentration and centralization of production . . . to be attained not through the detours of market effects. The change in national economic structures . . . is most efficiently realised by the state plan. The importance of common planning as the

> key link in the economic mechanism must be improved in every possible way. I would not consider the realization of market laws . . . to be the most important action we must take to accelerate the process of economic integration.

Bogomolov was reflecting the position taken by Kosygin (1974, pp. 430–57) in his speech to the 23rd (Special) CMEA Session held in April 1969, which is considered by many western and eastern commentators to have initiated the current round of integration proposals. Kosygin firmly indicated the supremacy of planning levers over monetary measures: 'The Planning Principle must be the basis of co-operation of Socialist Countries. Through this a more effective use of monetary levers will lead not to a lessening of the role of planning, but to its improvement.'

The methods proposed by Kosygin to achieve integration were remarkably similar to those proposed by Khrushchev in 1962 but carefully avoided the creation of any supranational agencies, 'substituting for national planning organs, ministries or departments'. The initiative in formulating joint plans was to be taken by the national bodies themselves. The methodology to be applied was that

> Plan Co-ordination should commence with joint activity in formulating long-term forecasts for economic development which will subsequently form the basis for co-operation in formulating members' five year plans and will finally be introduced into annual plans'. [The CMEA was to stimulate] a tighter drawing together [of CMEA] with national planning organs and of national planning organs with one another.

The economic methods were almost identical to those proposed by Khrushchev, including:

> Improved joint planning activity in certain branches of production of joint interest to members as a whole, involving the calculation of balances of production and consumption of specific commodities, especially raw materials . . . the joint financing of construction projects and the establishment of a Joint Investment Bank to provide medium and long-term credits for capital construction.

The rejection of market principles by Soviet authorities and the shift back towards joint planning (which appears to reflect a 'Soviet Gosplan' view) was probably influenced by internal political circumstances, although events in Czechoslovakia strengthened the opposition to market-type reforms.

The proposals for commodity convertibility and free turnover did present the USSR with a threat of strategic as well as purely ideological significance. A Hungarian-Czechoslovak agreement permitting non-quota turnover based on direct inter-enterprise contacts took effect from 1 January 1968. The open interest shown in these proposals by Polish economists, if carried to fruition, would have presented the USSR with a group of contiguous states spreading the length of central Europe, linked by principles of market socialism. This would have implied the development of a customs union every bit as strong as those vetoed by Stalin. Furthermore, direct links could have been extended to enterprises in Yugoslavia and possibly even the West.

The Warsaw Treaty Organisation's invasion of Czechoslovakia, and the subsequent recentralisation of the economy, removed much of the impetus to this form of integration, while the enunciation of the Brezhnev Doctrine provided an ideological justification for the preservation of the status quo.

The Planned Approach to Integration

Although the disagreement concerning the degree to which monetary levers should be subordinated to planning levers had not been fully reconciled by the time of the endorsement of the 'Complex Programme of Socialist Integration' in 1971, the measures that have been taken to implement the Complex Programme in the 1970s place greater emphasis on planning rather than market measures, including:[2]

(i) The establishment of the CMEA Committee for Co-operation in Planning Activity (1972) composed of the heads of the State Planning Agencies, and the proposal to establish a similar body for Co-operation in Material-Supply Allocation at the 28th (1974) Session.

(ii) The decision of the 27th (1973) Session that special sections for integration measures were to be introduced into members' annual, five-year and long-term plans.

(iii) The Agreed Plan for Multilateral Integration Measures for 1976 to 1980 adopted at the 29th (1975) Session.

(iv) The long-term Target Programmes for Integration Measures (1975 to 1990) approved at the 30th (1976) Session.

(v) The establishment of the International Investment Bank in 1971 and its subsequent operation, which has primarily been to promote joint planning measures.

The nature of this approach, and the inter-relationships between the various measures, can best be illustrated by the joint construction programmes included in the Agreed Plan of Multilateral Integration Measures for 1976 to 1980. Although the major motive force behind the Agreed Plan appears to have been the Soviet Gosplan, and the majority of initiatives have emanated from Soviet delegations to the CMEA, the CMEA Committee for Co-operation in Planning is responsible for their execution on the basis of the interested party principle. Decisions of the Committee do not apply to members who have declared their non-interest in specific programmes.

The major concrete proposals included in the Agreed Plan involved co-operation in the planning and pooling of resources for the joint-construction of ten projects to an estimated capital value of 9 milliard roubles. Ninety per cent of the expenditure concerned the development of fuel and raw materials for bloc use and eight of the projects were situated in the USSR. The majority of the projects therefore provide for the joint participation of the CMEA countries in the development and transportation of Soviet raw materials for consumption by participating countries, who will provide machinery and equipment, construction materials, labour and financial resources, to be repaid in products over a period of twelve years. The role of the International Investment Bank and the subservient role of monetary measures to plan instructions can be seen from the methods by which joint construction projects are drawn up.

The total cost of the project is initially estimated on the basis of internal wholesale prices in the country of construction, and is subsequently converted through foreign trade prices into transferable roubles, to arrive at the CMEA cost of the project. The proportionate participation of the creditor nations is then determined by their share of the output of the project on completion. The creditor nations (ie, those participating in the project) advance credit through the International Investment Bank to the debtor nation (ie, the country in which construction takes place) to cover the cost of their participation. Subsequently, the debtor nation issues to the creditor a detailed nomenclature of supplies and equipment for the construction of the project to be supplied against that credit. After a period of bilateral negotiations, the final details, including terms of delivery, are specified in foreign trade contracts and are written into members' national plans through the 'special sections', which are harmonised with the relevant operational plans (Lebedinskas, 1974, pp. 76–80).

In many cases, the International Investment Bank has raised convertible currencies to finance joint construction projects, thereby introducing an element of international commodity convertibility and flexibility into joint construction proposals (ie, the recipient of credits granted in convertible currency can use the funds to obtain any commodity it wishes from outside the bloc). Where a creditor country's contribution is in the form of convertible currency, the content is converted into transferable roubles, and is repaid by an equivalent value of commodities resulting from the project (Pecsi, 1981, p. 183).

There are clear signs that the momentum towards improved co-ordination of planning has slowed down in the late 1970s. Further Agreed Investment Plans were to be concluded for 1981 to 1985 and for 1986 to 1990 on the basis of the Long-Term Target Programmes for five major sectors of the economy — fuel, energy and raw materials; machine-building; agriculture; transport; industrial consumer goods. However, the volume of investment agreed for 1981 to 1985 is minimal. This contrasts with the expectations of some Soviet economists who had envisaged that the long-term target programmes would be extended beyond a few basic construction projects, to involve the examination of investment needs in entire industrial sectors from a bloc rather than a national viewpoint, and to the allocation of investment projects to specific countries with the establishment of sanctions to be taken against members who do not observe the agreements. Furthermore, they proposed that specific responsibility for the execution of the programmes would be given to international agencies, who would exercise exclusive control over resources (Promsky, 1977). The supranational implications of these proposals are clear and were resisted by Romania at the 1978 CMEA Session, and have also been viewed by Hungarian economists as impractical at the current time. Pecsi (1981, p. 77), for example, argues that although 'integrated community wide planning will eventually be necessary, limited investment funds and domestic constraints now permit only limited supranational planning even through target programmes'.

In the second half of the 1970s the attention of economists in Hungary, and to a lesser extent Czechoslovakia, returned to models of market integration. Vincze (1978), Kiss (1976), Pecsi (1981) (Hungary) and Chvojka (1977) (Czechoslovakia) have proposed models that include decentralisation of foreign trade decisions to the enterprise level, involving direct contacts between enterprises in different countries, and including both currency and commodity convertibility. The political implications of these proposals have been made clear by their authors. Chvojka (1977, pp. 20–1) admits that his proposals are incompatible with Soviet methods of planning 'convertibility is not compatible with an excessively command type, detailed physical kind of planning.'

Vincze (1978) proposes to

> increase the role of money so that it plays the role of a general means of purchase . . . [but to achieve this] money has to play an active role in the national economies of the individual countries . . . production, trade and consumption of products should not be regulated in the economic control systems of member countries by instructions expressed in physical units . . . the systems of economic control in individual countries should make it possible for the economic units (enterprises, trusts, farms) to buy commodities freely from abroad as well as sell products abroad . . . they should not be tied by compulsory prescriptions regarding the sale and purchase of products.

Other Hungarian economists have argued that such proposals are politically impracticable — for example, Bartha (1976) argues:

> The development of direct relationships among enterprises relying on a profit motive assumes changes in the methods of internal control of the economy that are impossible to implement without completely changing the system of control of the whole economy. And this — as is well known . . . is not on the agenda of the socialist countries.

The Impact of the Energy Crisis on East European Economic Relations with the USSR

There can now be little doubt that the Middle-East crisis and resulting oil price increases of 1973–4 provided a major impetus to the joint construction projects undertaken in the Agreed Plan. Although two projects, the Kiembayev Asbestos Combine and the Ust Ilimsk Cellulose Project (both situated in the USSR) were agreed before 1973, the majority of the construction projects agreed subsequently involved the extraction and transportation of energy or the extraction and processing of basic raw materials. Several reasons can be advanced for this. On the demand side, the increased price of alternative supplies of energy outside the bloc increased the potential rate of return, and therefore the attractiveness of investment in bloc energy sources. On the supply side, fuel and raw materials outputs can be expressed more easily in physical units than more complex products, and are therefore better suited to solutions based on physical planning. Furthermore, the size and nature of projects such as the Orenburg gas pipeline meant that they could, at least in theory, be divided more easily into identical physical components, without requiring such accurate value comparisons as would be required for more complex construction projects. In the case of the Orenburg pipeline, Hungary, the GDR, Poland, Czechoslovakia and Bulgaria were each to be responsible for the construction of just over 500 kilometres of pipe for which they would receive 2.8 billion cubic metres of natural gas per annum. Romania agreed to provide foreign currency for the purchase of equipment from the West and will receive 1.5 billion cubic metres (Bethkenhagen, 1977, p. 48).

The initial net impact of the 1973 increases in world energy prices on the terms of trade was a considerable boost to the USSR as a substantial exporter of oil and oil products; was approximately neutral to Romania (which was a marginal net exporter of oil and oil products until 1977) and to Poland (whose coal exports increased in value); but was a severe disadvantage to Czechoslovakia, Hungary, the GDR and Bulgaria. By the end of the decade the collapse of Polish coal production and Romanian oil output and the failure of domestic energy output to meet plan targets resulted in those countries requiring substantial energy imports, and therefore being considerably disadvantaged by the second round of world oil prices increases in 1979.

The introduction of the sliding world average price system involved both a considerable Soviet subsidy to the East European nations and a considerable increase in the cost of oil purchases. Estimates of the division of these costs and benefits between the

Table 9.1: Soviet Trade With Eastern Europe After The Oil Crisis (billion transferable roubles per annum)

	1974	1975	1976	1977	1978	1979	1980
A. Basic trade data							
Soviet exports to East Europe	8.7	11.9	13.1	15.3	16.9	18.5	20.9
of which: Energy	1.6	3.2	3.7	4.7	5.6	7.0	8.5
Oil	1.1	2.1	2.5	3.3	4.2	5.0	5.9
Gas	0.1	0.3	0.4	0.6	0.7	1.1	1.7
Soviet imports from East Europe	8.6	11.3	12.2	13.9	16.8	17.5	19.1
of which: Machinery	4.0	5.2	5.7	6.6	9.2	9.3	9.8
B. Estimates relating to oil price changes							
(i) CMEA price as % of world oil price	28%	60%	58%	65%	85%	82%	51%
(ii) Soviet oil subsidy to East Europe (billion roubles)	2.4	1.3	1.6	1.6	0.6	1.0	4.5
C. Soviet gains over 1974 (billion roubles)							
(i) Claims due to CMEA oil price rise		1.0	1.4	2.0	2.6	3.3	3.6
(ii) Claims due to net barter terms of trade		0.8	1.4	2.1	2.6	3.0	3.4
(iii) Net resource transfer		0.3	0.6	0.7	2.4	1.9	1.6
Balance of trade surplus	0.1	0.6	0.9	1.4	0.1	1.0	1.8

Sources: Section A, *Vneshnyaya Torgovliya za SSSR*, various years; Section B (i), J. P. Stern (1982), p. 26; Section C, based on calculations for: Smith (1981a) extended to include estimates for 1980.

USSR and the East European nations are shown in Table 9.1

The value of the Soviet oil subsidy to Eastern Europe has been estimated on the basis of the difference between the world market price for crude oil and the intra-CMEA crude oil price, estimated according to the sliding world average formula, multiplied by estimates of quota deliveries. (Soviet deliveries above agreed quotas are normally paid for in hard currencies at current world market prices.) These estimates indicate that in 1974, before the price basis was changed, the annual Soviet implicit subsidy to Eastern Europe was 2.4 billion roubles (Table 9.1, row 8). The new formulas reduced the subsidy to 1.3 and 1.6 billion roubles in 1975 and 1976 respectively, and as the full impact of world price increases was fed into the CMEA price formula, the subsidy was gradually eliminated and had fallen to 0.6 billion roubles in 1978. The second round of Middle East price increases, in late 1979, substantially increased the implicit subsidy to 4.5 billion roubles in 1980, and a similar process of gradual elimination should take place in the early 1980s. The total subsidy amounts to 13 billion roubles from 1974 to 1980. Substantial price rises for oil exports have occurred, however.

The value of Soviet energy exports to Eastern Europe had increased by 6.9 billion roubles in 1980 over the 1974 level — 4.8 billion roubles was accounted for by exports of oil and oil products, of which approximately 3.6 billion roubles can be attributed to price increases (after making a fairly generous allowance for increased volume and deliveries of oil products). The sum of oil price increases over 1974 price levels on Soviet deliveries to Eastern Europe amounts to 13.9 billion roubles from 1975 to 1980.

This price increase could be balanced in Soviet-East European trade in any combination of three possible forms:

(i) The USSR could increase the price it pays for imports from Eastern Europe.

(ii) The USSR could run a trade surplus with Eastern Europe and therefore effectively extend credit for a determinate or indeterminate period.

(iii) The USSR could require the East European countries to make increased deliveries of goods and services.

In an earlier publication (Smith, 1981a), I attempted to estimate this breakdown by calculating an implicit price and volume index for Soviet trade with Eastern Europe. These data have been summarised in Table 9.1 and extended to include the effect of the

1979 price increases. Some of the more tentative conclusions can now be strengthened as more data have been made available.

Estimates of the annual net gain to the USSR arising from the increased price of Soviet exports to Eastern Europe, minus the increased price of Soviet imports from Eastern Europe (Table 9.1, row 10) are very similar to the gains arising from oil price rises alone. Consequently, the first possibility that the USSR could compensate for increased *oil* prices by increasing the price paid for imports from East Europe accounts for only 0.6 billion roubles, although price increases for *other* Soviet exports, including raw materials and natural gas, have been offset by an increased price paid to East Europe for exports to the USSR.

The Soviet Union has run visible trade surpluses with the East European countries throughout the period. These were quite substantial from 1975 to 1977 but were virtually eliminated in 1978, and then reopened in 1979 and 1980. The surplus amounts to 5.8 billion roubles from 1975 to 1980 (or 41 per cent of the increase resulting from oil price gains).

Finally, an increased net flow in the delivery of commodities amounting to 7.5 billion roubles has been made to pay for oil price increases. This amounts to 54 per cent of the gain arising to the USSR as a result of the changed price formula for oil.

The composition and timing of this resource flow from Eastern Europe is revealing. From 1974 to 1977 the net flow of resources from Eastern Europe to the USSR was only marginally above the 1974 level. In 1978 the Soviet trade surplus was virtually eliminated as Soviet imports from Eastern Europe increased by 2.9 billion roubles (2.3 billion in real terms) over the 1977 level, as a result of which the real net flow of resources to the USSR was 2.4 billion roubles higher than the 1974 level. Although the net flow of resources from Eastern Europe declined slightly in 1979 and 1980, this decline can be attributed entirely to Polish trade deficits with the USSR, while the other East European countries have continued to make a resource transfer at around the 1978 levels but below that required by the increases in 1978 and 1979 CMEA prices, resulting in further balance of payments deficits.

The increase in East European exports to the USSR is concentrated in deliveries of machinery and equipment, the nature of which is not specified in Soviet statistics. Increases in the deliveries of unspecified machinery and equipment are approximately equally distributed between all the East European countries with the exception of Romania, which receives no quota deliveries of Soviet oil, and does not participate in the majority of joint construction projects. The size of the amount concerned and its distribution makes it highly probable that the increased flow of resources from Eastern Europe to the USSR is accounted for by the joint construction projects.

It is also noticeable that the concentration of deliveries (which come to approximately $100 per worker per annum in the five countries concerned) in the second half of the five-year plan, coincides with the fall in the rate of growth of real retail turnover in each of the countries concerned, with no commensurate decrease in the growth of wage rates, culminating in open inflation.

The joint construction projects should not be equated with Soviet exploitation of its satellites. In the long run the participating countries will receive raw materials and energy sources as a result of their capital participation, while the sliding world-average price system will imply a considerable subsidy to the East European countries, if world oil prices do not stabilise. Furthermore, by participating in such schemes to ensure the supply of raw materials within the bloc, CMEA countries should be enabled to maintain a higher level of aggregate demand and supply than would result from an attempt to maintain external equilibrium through deflationary policies.

The projects will, however, impose considerable costs on the East European countries, and current disagreements over these issues appear to have prevented any substantial new projects being developed. Current East European complaints include the argument that too much of the construction cost involves developing Soviet infrastructure, and that participating countries have little or no control over the calculations made by the host country to measure contributions (Pecsi, 1981, pp. 72–3). Furthermore, the prices of the resulting outputs are to be revised according to the moving-average price system. Consequently, even though physical quantities are specified in the agreements, the real rate of return to participants is uncertain, and the prospect may be that participants will be required to make further outgoings, to balance bilaterally any unforeseen price increases.

A more critical issue in the 1980s will be the terms on which the USSR supplies oil and natural gas to the East European countries. The USSR has agreed to supply Eastern Europe with 80 million tons of oil per annum (the contracted level for 1980) for the period from 1981 to 1985 on the basis of the world moving-price system, for which it will be reimbursed in products. Any deliveries above this amount will be subject to separate, bilateral negotiations and the USSR may well request payment in hard currencies at existing world market prices.

If world market prices should stabilise at their 1982 levels, the subsidy element involved in Soviet oil deliveries would be gradually eliminated by 1985, while the East European countries would by then be paying 5–6 million roubles more per annum for their oil than in 1980, a sum that would absorb much, if not all of their potential growth in that period. In many cases, the USSR may find itself forced to seek specific methods to alleviate the pressures this will cause, either by continuing to accept soft goods, or by allowing the East European countries to continue to run balance of payments deficits at a substantially higher level than those incurred in 1979 and 1980. Soviet room for manoeuvre will, however, be limited by the low living standards of its own population relative to those of Eastern Europe, and the cost of development programmes in Vietnam, Cuba and Mongolia.

The alternative for the East European countries of purchasing oil from Middle-East countries would, at 1982 consumption levels and world prices, cost the East European countries 20 million dollars per annum, and would present a far greater burden to the East European countries, as can be seen from the case of Romania, which is discussed in the next chapter.

Notes

1. For a discussion of the circumstances surrounding the formation of the CMEA see van Brabant (1980, Chapter 1) and Kaser (1967, Chapter 2).
2. For a more detailed discussion see Smith (1979b).

10 EAST-WEST TRADE AND THE STRATEGY OF IMPORT-LED GROWTH

Techology Transfer and Economic Growth in Eastern Europe

The impact of the transfer of technology from technically advanced countries to technically less advanced countries on the rate of economic growth in the latter is not immediately obvious. Clearly, if by some process the most efficient technology could be effectively transferred throughout the nations of the world, so that the productivity of every worker in the world were raised to the highest international level pertaining in that industry, the growth rate of labour productivity would be highest in those countries in which the initial levels of labour productivity were lowest.

A number of factors, including educational levels and industrial training, cultural traditions, the level of infrastructure and the nature of the economic and social system, may inhibit the transfer of innovations from one country to another, and may prevent the efficient application of innovations so transferred. As the factors that prevent the efficient application of new innovations and their diffusion through the economy are likely to be the same as those that cause a low initial level of labour productivity, it is not clear whether the possibilities of improving labour productivity through technology transfer will be positively or negatively related to the absolute level of labour productivity, or whether a policy of the deliberate acquisition of technology will be effective.

Stanislaw Gomulka has undertaken an exhaustive series of studies of this problem, which indicate that the successful transfer of technology provides a principal explanation of differences in the rate of economic growth in general and of growth rates in Eastern Europe in particular.

Gomulka's major study (1971), embracing 55 nations ranging in per capita GDP from the USA to Burma, indicated a pyramid or 'hat-shaped' relationship between the absolute level of output per worker and the rate of growth of labour productivity. Countries with a very low level of output per worker (which Gomulka equates with a low level of technical development) also experience low rates of growth of labour productivity, and are unable to benefit significantly from innovations taking place in the technically more advanced industrial sectors of the world. Countries with an intermediate level of labour productivity and technology experience the highest growth rates of labour productivity, as they are capable of absorbing a large number of technical innovations from the technologically more advanced countries, and raise their labour productivity accordingly. Countries with the highest absolute level of output per worker and technical development experience relatively low rates of growth of labour productivity, as their rate of growth is constrained by the level of domestic innovation and those innovations that can be absorbed from technically advanced sectors in other countries.

Countries at an intermediate level of development may grow more quickly than the technologically most advanced nations of the world, but if and when they approach the levels of the advanced countries their growth rates will inevitably slow down, and they will only overtake the technological leaders if they are capable of generating a higher level of innovative activity domestically.

This process was reflected in the USSR according to Gomulka (1976) by an exceedingly high rate of industrial growth in the immediate post-war period, resulting from the re-equipment of war-damaged factories and the absorption of technical innovations from which the USSR had been isolated during the war period. Similarly, the East European countries were capable of generating a rapid rate of industrial growth in the early 1950s as they repaired their war-damaged economies, embarked on the process of rapid industrialisation, and absorbed technical innovations which had remained undeveloped both during the war and in the case of the less industrialised economies, in the inter-war period. Gomulka (1982b) has also shown that the rate of growth of industrial output in Eastern Europe in the late 1950s and early 1960s was highest in those countries with the lowest levels of per capita income (Bulgaria and Romania), was lowest in the countries with the highest levels of per capita income (Czechoslovakia and the GDR), while intermediate rates of growth were recorded by those countries with medium income levels (Hungary and Poland).

This is partly a reflection of the policy of extensive growth, as the rate of growth of both the industrial capital stock and the industrial labour force was highest in the more agrarian countries, as priority was given to industrial development and more labour was transferred from agriculture to industry. Gomulka also shows that the rate of growth of output per worker *in industry* was higher in those countries where the absolute level of output per worker was lower (although the GDR appears to grow somewhat more quickly than the general rule would indicate).

This cannot be attributed to pure extensive growth (ie, replicating existing technology) which would not result in a growth rate of industrial output per worker, and therefore would not result in differences in the rate of growth of output per industrial worker between countries. The newly created capital stock must therefore have embodied production techniques (including those involving economies of scale) that were not present in the old capital stock, and these developments must be greater in the case of the less industrialised countries than the more industrialised. Further tests by Gomulka indicate that a significant element of the difference in the rate of growth of industrial productivity can be attributed to differences in the level of labour productivity, from which it may be inferred that technical transfer within the bloc from the more industrialised to the less industrialised East European countries was a significant factor in the industrial growth rates of the latter in the 1950s.

The ability of the more industrialised countries to augment their growth rates of labour productivity, by acquiring technology from more advanced countries outside the bloc, was considerably hampered by CoCom restrictions on the export of commodities embodying technology from the West to the East, while the GDR increased its isolation from western ideas by building the Berlin Wall in 1961. As a result, the more advanced countries were becoming increasingly dependent on domestic innovation to stimulate their growth rates.

This analysis has implications for the process of CMEA integration and the dispute between the industrialised and agrarian nations within the CMEA. The policy of specialisation in research and development, advocated by the more industrialised nations, would have allowed them to concentrate their efforts on raising technology in the more advanced sectors of industry, while leaving the less developed to concentrate on other areas. In those areas where production was to be continued in more than one country of the bloc, the concentration of the research and development effort would allow the less developed countries to gain a technical lead in specific sectors, which would permit technical transfer to take place from the lower income to the higher income economies, as well as in the other direction.

On the other hand, the concentration of production of high-technology products in the more industrial nations would remove a major source of technical transfer to the less developed countries and could therefore reduce their levels of domestic production if not consumption. As a result, Romania argued that specialisation should not take place until the differences in the levels of development between the countries had been substantially reduced, and investigated the prospects for acquiring technology directly from the West.

Methods of Technology Transfer

A distinction is frequently made in the West between 'negotiable' and 'non-negotiable' technical transfer. Hanson (1981a, p. 15) defines negotiable transfers as those that can be influenced by the state of relations between governments and can, in particular, be influenced by the policies of the supplying governments. Such transfers, which include high-level and continuing contacts between scientific organisations, large-scale sales of capital equipment including the issue of export licences and government backed credit and export guarantees, involve a considerable degree of personal contact and bargaining. It was this channel of technical transfer that had been largely closed by the actions of CoCom on the western side and by the restrictions on foreign travel, hostility towards foreign investment, and so on, on the eastern side, and the low level of West-East emigration.

Consequently, West-East transfer of technology had been largely restricted to non-negotiable transfers. These are defined (Hanson, 1981a, p. 15) as policies engaged in by the recipient country to obtain information about technical developments in other countries which involve a far lower degree of personal contact, including the study of scientific and technical journals, the purchase of single items of machinery for the purpose of imitation (reverse engineering), and observation of processes in other countries, including espionage.

A further distinction could also be made between the nature of technical transfer and its relation to the process of production, ranging from the dissemination of 'pure' knowledge across international boundaries, through the dissemination of knowledge and information concerning production processes, to the transfer of actually operating production processes and techniques embodied in machinery and equipment.

Non-negotiable transfer is probably better suited to the spread of pure knowledge and certain aspects of information concerning production processes (industrial espionage, the scrutiny of trade journals and applied-science journals), while negotiable transfer is required for the transfer of many aspects of information concerning production processes (the purchase of licences, patents, and so on) and for the transfer of actually operating production processes themselves. The low level of personal contact involved in non-negotiable transfer also prevents any discussion of problems that may arise in the process of adapting imported technology to domestic conditions, and even of problems that the initial developer of the process may have had in implementing the process.

Hanson argues (1981a, pp. 49–80) that the USSR does exhibit a systemic bias against technical developments (which would also be applicable to the East European CPEs), but that this bias does not occur so much at the level of pure research, but is caused by the separation of research and development and by the lack of incentive to enterprises to bring existing knowledge into series production, to develop new products, and to diffuse new production processes rapidly through the economy. The low levels of negotiable transfer therefore reinforce a principal area of systemic weakness, inherent in the structure of CPEs. The solution of this problem domestically would require changes in the operation of the economic system, while the acquisition of foreign technology through negotiable transfers would require changes in the state of East-West relations, extending beyond relations at governmental level to involve changes in the attitude to commercial relations with western enterprises.

Dunning (1969, pp. 246–8) argues that the international transfer of technology, including managerial expertise, was a major source of economic growth in Western Europe in the 1960s, and that this was largely facilitated by direct investment by multinational corporations based in the USA, which maintain a relatively high expenditure on research and development and considerable growth potential.

The conventional explanation for this phenomenon is the product-cycle model. This theory proposes that R & D is a fixed cost and is therefore more profitable for companies based in economies with large domestic markets, which can guarantee a high volume of sales for new products without the fear of the imposition of trade barriers designed to protect producers adversely affected by the new product or process.[1] The USA offers the largest domestic market in the West, both in terms of population size and purchasing power, and in addition many larger corporations, particularly in the field of aviation and electronics, undertake defence-sponsored research, much of which has spillovers into normal commercial use.

In the initial stage of innovation (ie, the first stage of commercial production) the innovating enterprise will enjoy a monopoly which will enable it to obtain a high selling price for its product and may make cost factors relatively unimportant, enhancing the value of a high income but (high labour cost) base for production. The advantages of a domestic production base will be increased where proximity to local technologists is important in the debugging stage of production, where feedback from sales departments to production is important, and may be essential on security grounds when spin-off from the defence industry is involved.

As initial production problems are overcome and the new process or product reaches full production, the desire for sales volume will require price reductions. Furthermore, competitors who have been adversely affected by the innovation may be forced to retaliate by bringing similar, but not identical products (and therefore not covered by existing patents) on to the market (eg, the proliferation of different types of calculators, electrical goods, small cars, training shoes).

At this stage, competitive pressures may lead to the need to seek new markets for the product and to the desire to reduce production costs. This can result in the transfer of production to countries with lower unit labour costs, but with a technically competent workforce, with the intention of both establishing markets overseas and importing the resulting output from the foreign subsidiary. Alternatively, the corporation may attempt to recoup some of its initial R & D costs by selling licences overseas, which may involve less risk of establishing domestic competition than a domestic licence sale, while concentrating its own research on the next generation of production (Hanson, 1981a, p. 29).

The product-cycle model does not provide a total explanation of the process of technical transfer from the more advanced industrial economies to economies at an intermediate stage of development which will hold good for all periods of time either in the past or the future, but it does provide an explanation of the channels of technical transfer in the post-war era, from which the East European countries were isolated by the nature of their political as well as their economic system.

Furthermore, the technical leaders in the bloc, the GDR and Czechoslovakia, possessed small domestic markets, while the system of annual bilateral balancing in intra-CMEA trade provided no guarantee that new products (particularly intermediate goods) would be accepted by CMEA partners, or that satisfactory products would be offered in exchange, thereby substantially reducing the returns to domestic R & D. The USSR alone possessed a sufficiently large domestic market to absorb new commodities in exchange for energy and raw materials, but the absence of competitive pressures in this market in the 1950s reduced the incentive to apply new processes and develop new products.

Finally, although the USSR has a large military and space sector capable of generating technical innovations, security fears prevent many of the innovations in this sector from being transferred to other sectors of domestic production, and in particular being transferred to Eastern Europe.

Soviet and East European Attitudes to West-East Technology Transfer

Soviet Perceptions

Soviet acceptance of the possible benefits to be obtained through co-operation with western corporations originated in the Khrushchev era. The speech in which Khrushchev (1962b) proposed a common planning organ also contains many positive references to the organisational and managerial activities of multinational corporations and their role in integration in Western Europe. Hanson (1981a, p. 108) shows that Soviet co-operation with multinationals involving long-term contacts, including the training of engineers, had begun in the chemical industry in the Khrushchev era.

The scale and importance of Soviet industrial contacts with the West increased in the mid-1960s, including the Fiat-Tolyatti deal, which introduced several western managerial practices, and included an arrangement for the export of 30 per cent of the output (Hanson, 1981a, p. 109). A difference of opinion appears to have developed between Kosygin, who favoured the extension of contacts with the West for the acquisition of technology combined with domestic reform, and Brezhnev, who argued that the domestic capacity for innovation was underestimated. Ironically, as Brezhnev's power grew and attitudes towards domestic liberalisation hardened in and after 1968, Brezhnev appears to have become more aware of the limited innovative capacity of the unreformed system, and a positive policy towards the acquisition of western technology emerged in the shape of détente.

Technology Transfer in Eastern Europe and Economic Reform

Although the acquisition of western technology by East European countries in the 1970s has acted as a substitute for economic reform, market reformers regarded the two policies as complementary in the 1960s. In Czechoslovakia, Ota Šik (1971) argued that the continuation of trade links with the less developed CMEA economies would dampen domestic innovation, and proposed to expand trade links with the West, culminating in an attempt to seek a 500 million dollar loan from West German banks to purchase western machinery and equipment.

Price reform was seen as a necessary adjunct to this policy in order to facilitate the comparison between imported and domestic technology, and to accurately assess the costs of related materials and components (Sik, 1967, pp. 91–2). Decentralisation of decision-making was also considered necessary in order to provide enterprises with the incentives to adopt new technology and for the diffusion of new technology through the economy. Finally, in order to generate a sufficient volume of hard currency exports to pay for imports, it was argued that enterprises would have to be empowered (and encouraged) to respond quickly and flexibly to changing conditions on world markets, and to produce commodities of world quality specifications.

The logical implications of this approach, which were clearly explained by Ota Sik on Czechoslovak television in 1968,[2] were the abolition of the centralised foreign trade monopoly and the *Preisausgleich* in order to bring enterprises into direct contact with foreign market and price conditions, the possibility of bankruptcy of enterprises 'innovated against' and a virtual abandonment of the 'no frictional unemployment' constraint.

Similarly, the architects of the Hungarian New Economic Mechanism proposed a direct link between domestic reform and imported technology. Enterprise profitability was a necessary measure to stimulate domestic innovative activity; low quality production was to be eliminated by imported western technology and new export-oriented industries were to be created. Direct planning procedures were to be dismantled and enterprises were to be empowered to retain profits for reinvestment (Abonyi, 1981, p. 137).

Intriguingly, however, the most drastic initial increase in machinery imports from the West was undertaken by Bulgaria in 1966 and 1967, undoubtedly with Soviet approval or possibly even as a direct Soviet experiment.

From December 1965 until July 1968, Bulgaria pursued an economic reform similar to (but slightly more extensive than) that implemented in the USSR, involving the concentration of enterprises into associations and a reduction in the number of compulsory enterprise indicators (Vogel, 1975). Simultaneously, Bulgarian imports of machinery and equipment from the West doubled from 71 million dollars in 1965 to over 150 million dollars in both 1966 and 1967. It was hoped that the creation of foreign trade enterprises would stimulate direct links with foreign enterprises, which would increase the competitiveness of Bulgarian industry in western markets and, in particular, that the creation of agro-industrial complexes would introduce a greater use of scientific methods in agriculture which would result in increased output of products that could be marketed in the West.

Bulgarian policy from 1965 to 1968 exhibits three trends: (1) the creation of large-scale units, combined with attempts to introduce new technology; (2) an economic reform giving greater powers to enterprises and in particular attempting to stimulate export efficiency; (3) the import of western technology.

The reforms had hardly had time to take effect or be evaluated when a recentralisation of industry was announced in July 1968 which placed greater emphasis on administrative reforms. Vogel (1975, p. 208) argues that no domestic policy motive can be found for the reversal of the reforms. Imports of western machinery declined to 87 million dollars in 1969 and only started to recommence their growth under the influence of détente in the early 1970s.

Romania embarked on a policy of sustained import-led growth

in 1967. In the 1950s Romania conducted a small volume of trade with the West which consisted mainly of exports of oil, oil products and foodstuffs, and imports of basic manufactured goods and machinery and equipment. Romania's per capita trade with the West in 1960 was equal to that of Bulgaria; making them jointly the lowest in the bloc.

The construction of the Galati steel-mill required increased imports of machinery and equipment from the West which amounted in total to 560 million dollars in the 1961–5 plan period. Imports from each CMEA country were considerably curtailed in 1965 following Ceausescu's accession to power, the majority of the reduction being concentrated in machinery and equipment and raw materials.

Romania did not substantially increase her imports of machinery and equipment from the West until 1967, when they amounted to 360 million dollars, 169 million of which came from West Germany. (In the same year Hungarian and Czechoslovak imports from the West amounted to 112 million dollars and 142 million dollars respectively.) This policy was accompanied by significant political gestures to the West, including the recognition of West Germany and a neutral stance on the Arab-Israeli war.

Two different schools of thought were apparent amongst Romanian economists and political leaders concerning the relationship between import-led growth and domestic economic reform. Reform experiments were conducted before 1967 and a compromise set of reform proposals containing some decentralist elements were approved by the Party Congress in 1967, but were shelved in 1968 as a result of anxiety over Czechoslovakia, and domestic reforms were limited to administrative measures.

Imports of western machinery and equipment were also reduced in 1968, but remained the highest in the bloc (approximately 300 million dollars per annum) until 1972, from which date Poland became the largest East European importer of western machinery and equipment.

The Strategy of Import-Led Growth in the 1970s

It appears therefore that a discreet Soviet trial of the policy of import-led growth, involving a rapid influx of western machinery and equipment combined with limited domestic reforms and specific measures to improve the competitiveness of export industries, was first launched in Bulgaria in 1966 and 1967, but was cut back in 1968 when the political repercussions of the reform movement in Czechoslovakia became apparent. Further examination of the effects of co-operation with western corporations, and with Fiat in particular, continued in the USSR from 1968 to 1971, by which time the Politbureau had been convinced of the benefits of this and other aspects of improved relations with the West, and the policy of détente was launched. How the USSR appraised Romanian developments is not clear.

The economic aspects of détente included the rapid acquisition of western technology on the basis of the establishment of co-operation ventures with western multinationals and large-scale imports of western machinery and equipment, largely on credit. A distinction may be made between the (intended) impact of the policy on the USSR and the East European countries. Although the build-up of machinery and equipment imports was rapid in both areas — according to western data Soviet imports of machinery and equipment grew from just under 1 billion dollars in 1971 to 4.2 billion dollars in 1975, while the corresponding figures for East Europe were from 1.5 billion dollars to nearly 5 billion dollars — the sheer size of the USSR and the low level of dependence on imports in general meant that this only accounted for about 4 per cent of total Soviet investment in the period. In the smaller East European countries the quantitative importance of western machinery was far greater, reaching to nearly 25 per cent of Hungarian investment in the 1971–5 plan period.[3]

A more significant difference in the strategy concerned the possibility of exports for hard currency and their impact on the balance of payments. The USSR possesses substantial resources of raw materials, precious metals and energy sources, which comprised a major part of their hard currency earnings. Following the US decision to demonetise gold in 1971, gold prices appreciated in world markets and the USSR, which had sold no gold between 1966 and 1970, sold a total of 920 tons between 1971 and 1975. Simultaneously, sales of other precious metals and diamonds were also increased, while even sales of arms for hard currencies were stepped up. Bad harvests in 1972 and 1974 and an apparent higher priority attached to domestic living standards resulted in purchases of grain from the USA and Canada, but as world oil prices rose in 1974 and 1975, and the value of domestic resources increased, the

USSR halved its sales of gold and covered grain purchases by borrowing in world markets. Much of the Soviet increase in indebtedness in 1975 must therefore be attached to grain purchases, and Soviet speculation in world markets, rather than to a deliberate policy of substantial borrowing to cover imports of machinery and equipment.

The position of the far less favourably endowed East European countries was significantly different. The acquisition of technology resulted in immediate balance of payments deficits as imports grew faster than the capacity to export. CMEA (CMEA, various years) statistics show that East European imports from the industrial West grew from 5 billion transferable roubles in 1971 to 12.5 billion in 1975, while exports only increased from 4.2 billion to 7.7 billion, resulting in a growth in the deficit on visible trade from 0.7 billion roubles (0.8 billion dollars) in 1971 to 4.8 billion roubles to 6.9 billion in 1975.

US National Defense Center estimates indicate that the net indebtedness of the East European CMEA nations rose from 4.6 billion dollars at the end of 1970 to 18.9 billion at the end of 1975 and to 56.3 billion dollars at the end of 1980 (see Table 11.2, p. 226). The growth of indebtedness, particularly in the period from 1971 to 1975, has varied between individual countries, largely in relation to the extent to which those countries embraced import-led growth, with Poland the most indebted with imports of western machinery and equipment of 5.6 billion dollars in the 1971–5 plan period and a net indebtedness of 7.4 billion dollars.

Co-operation Ventures: The Simulation of Multinational Investment

Balance of payments problems were also complicated by systemic factors. When a multinational invests in a host country it not only provides new techniques (knowledge capital) but if it is to retain ownership of the plant and equipment must provide finance for the project and markets for the outputs. In most cases the multinational may make a positive contribution to the host country's balance of payments by providing a short-term capital inflow (effectively buying domestic labour and equipment) and in the longer term by providing guaranteed export markets.

The company of course intends that the value of the resulting output of the plant (measured in the prices of the country in which it is sold) will be greater than the initial cost of the investment and will give a satisfactory rate of return. The initial investment should therefore add to the productive capacity of the host country and the profits to be transferred back to the investor should come from the surplus generated by the initial investment. If however the investment does not increase the output of the host country it is the original investor who, as risk taker, should bear the losses and the investor will have no financial claim against the host country other than the freedom to dispose of its own assets.

This is clearly an idealised description of the process of international investment, and even in cases where direct ownership and risk taking are carried by the investing company, the use of host government subsidies, export credits, international financial arrangements, transfer pricing, and so on, may make the net effect of any given investment far less certain to the host country even before the external effects of the investment on domestic producers are considered. Furthermore, it could be argued that it would be more profitable for the host country to construct the plant and market the outputs itself and bear the risks, taking the resulting profits and carrying any losses. This is effectively the case when an enterprise (or nation) undertakes licence purchase which may have considerable balance of payments risks for the purchasing country. Initially, the receiving enterprise must obtain and use foreign capital to purchase the licence, may have to import capital and machinery in the process of construction of the plant and subseqently import raw material and components for its operation and will then have to compete for sales in the domestic market, or, if it wishes to obtain foreign currency in overseas markets, possibly compete against the original licence holder who by that time may have improved the product or have imposed restrictions on the markets in which the products can be sold.

This was the problem encountered by the East European countries who were reluctant to allow western corporations to have full rights of ownership and disposal of assets in their territories, which, as well as posing ideological problems, would have conflicted with the principles of physical operational planning. Western companies, however, were unwilling to make the necessary financial outlays to construct plant and provide equipment when they had little direct control over the prices and quantities of inputs and outputs (ie, no commodity convertibility) and could not dispose of their assets if they wished to discontinue the project (eg, in the event of continued loss-making). Thus, although Romania and Hungary enacted legislation in 1971 to permit minority foreign equity holdings on their territories, the rights accorded to the investors were highly constrained (particularly in the case of Romania), consisted largely of profit-sharing and were insufficient to attract many investors.

The majority of East European countries attempted instead to imitate the process of multinational investment by establishing co-operation ventures on their territory which in principle involved the purchase of foreign technology, to be paid for by the sale of products resulting from the new plant.

In the purest form of this, the multinational delivers machinery and equipment, licences, documentation, and so on, on credit and receives in exchange products from the plant, specified in annual quotas, while the East European country retains full ownership of the plant and is responsible for the planning of inputs (including labour) and receives a share of the production. More diluted forms include subcontracting production, mainly of components, from the multinational to the East European country on the basis of direct instructions and documentation, without direct participation by the multinational in the construction of the plant. The East European country is effectively involved in selling domestic value-added to the western corporation but receives only a limited amount of technology. Another form of co-operation includes various types of co-production agreement whereby the multinational and the East European country specialise in the production of specific components which are subsequently exchanged and the finished products are assembled in both countries and are markedly jointly in third countries and/or separately in predesignated areas (Murgescu, 1967, pp. 67–74).

The first form of co-operation venture is the clearest attempt to imitate the product-cycle model and offers costs and benefits to both partners which are the simplest to analyse. The multinational provides the capital input and possibly arranges for its finance, and receives products in exchange. In terms of comparative advantage, the western partner is intended to supply capital-intensive inputs and the East European partner labour-intensive inputs. The East European partner benefits by receiving a technology which enables it to produce a complete product, while if the multinational is to receive the outputs of the plant, it may be more willing to pass on any subsequent innovations that may occur after the initial investment.

Western companies have frequently been unenthusiastic about such ventures due to their lack of control over the production and, in particular, the quality of outputs to be received. Furthermore, East European countries frequently evaluate labour costs at West European wage rates when negotiating the value of their own contributions to the project, even though the wages actually paid may be substantially lower. Consequently, many multinationals may prefer to establish wholly-owned subsidiaries in Third World countries, where the actual labour costs are lower, and a greater degree of product control may be exercised.

The more diluted forms of technical transfer, involving component production only, are less satisfactory to the East European countries as they receive only limited technical inputs (for example, for the assembly of computer software) which cannot be used to manufacture an entire product for domestic consumption. There may therefore be a considerable degree of ad hoc bargaining between the partners.

A Romanian agreement with Renault provides an example of a successful outcome to this process. Renault provided machinery and equipment on credit for the construction of a plant to manufacture gearboxes and axles and was repaid over a period of five years in the form of products which were used in the production of vehicles in France. Romania also received documentation for the production of Dacia cars for the home market, but now claims to export them to over 40 countries (Ionescu, 1974, pp. 114–15).

A major obstacle to the development of co-operation ventures has been that once an East European country has identified the technology most appropriate to its needs, it must attract a company or group of companies to enter into the co-operation venture. Large-scale construction projects in the West normally involve not only the corporations concerned in the production and sale of the finished product, but also specialist engineering and construction firms. In many cases where the production of the finished product itself did not require a licence or patent, or the multinational was not interested in receiving any of the outputs of the venture, the construction firms became the major partner in the project, and the pursuit of import-led growth compelled the East European countries to make large-scale purchases of machinery and equipment and complete installations (turnkey projects), and to make their own arrangements for marketing the finished products.

Large-scale capital projects in the West normally involve a considerable degree of credit and bank finance to cover the period before the venture starts to earn revenue. Where such projects are to be constructed overseas, the government of the supplying country may offer official government credit support, and certain loans and guarantees may be provided by the government of the receiving country.

The non-convertibility of the East European currencies makes loans provided in those currencies (eg, by the central bank or investment bank of the country concerned) unacceptable as a means of payment to western machinery manufacturers, who require payment in hard currencies to cover their own hard currency costs. Consequently, those East European countries which did not possess reserves of raw materials and precious metals which could be sold to raise hard currency, and were relatively inexperienced in selling manufactured commodities for hard currency, were forced to borrow hard currency from western banks to pay for imported machinery and equipment, and had to enter negotiations to obtain credits from western governments and/or obtain credits direct from the suppliers themselves.

The original intention of the import-led growth strategy was that after an initial spurt of borrowing the East European economies would be pushed into a dynamic growth path which would permit the loans to be repaid by the sale of outputs of the new technologically developed industries, or by ancillary developments stimulated by the original imports — for example, imported chemical plant would produce fertiliser which would in turn improve agricultural yields which, together with imported machinery for freezing and canning, would result in increases in the value of exports of processed foodstuffs (Lazaroiu, 1967).

Provided the imported technology generated a sufficient expansion of domestic output and exports, the volume of imported machinery could continue to rise to permit a permanent strategy of import-led growth to be maintained.

Gomulka (1982a) suggests that this did not necessarily require a surplus to be made in the hard currency balance of payments. If each individual project yielded sufficient hard currency earnings to pay off maturing loans and interest as they fell due, and western bankers retained confidence in the CMEA countries' ability to repay, a constant path of new loans, new growth and virtually permanent indebtedness could be maintained. Furthermore, the volume of machinery and equipment imported on credit could expand in proportion to the expansion of the domestic economy.

The Romanian case permits an evaluation of the second stage of the strategy, when imports and exports were to be brought on to a stable growth path, without the complicating factors of the oil crisis and western recession.

In the first stage of the rapid build-up of machinery and equipment imports lasting from 1967 to 1970, Romania's imports of machinery and equipment from the West (1.25 billion dollars) were roughly equalled by her deficits on visible trade with the West (1.1 billion dollars). In the second stage from 1971 to 1973, imports of machinery and equipment totalled 1.6 billion dollars, but visible trade deficits were reduced to a total of 0.5 billion dollars.

Romanian data measured in foreign exchange lei (which appreciated relative to the dollar and should eliminate the effect of western inflation) show that this was achieved mainly by a respectable growth of exports to industrial market economies of around 22 per cent per annum.

Data from Romania's trading partners indicates that fairly significant changes in the structure of Romanian exports had occurred and that Romania was now exporting 'medium technology' commodities. In 1965 85 per cent of Romania's exports to the EEC were composed of foodstuffs, wood and lumber, and refined oil products. By 1973 these items accounted for 48 per cent of Romania's exports, and the composition of Romanian food exports had shifted away from grains and cereals to meat and processed meat and fruit and vegetables (Smith, 1979a). The largest growth had taken place in clothing, textiles, furniture and footwear. The danger to Romania was that this labour-intensive export structure was vulnerable both to protectionist pressures and Third World competition, while the volume of Romanian indebtedness was being viewed as an obstacle to further credits in France and West Germany, which compelled Ceausescu to announce in July 1972 that the balance of payments would be equilibriated in 1974.

In fact, a surplus on trade with the West would have been achieved if the rate of growth of imports in 1972 and 1973 had been restricted to its 1967 to 1971 levels of about 6 per cent per annum. In practice, in 1972 Romanian imports from the West grew by 18 per cent and (after correcting for the effects of western inflation) machinery and equipment imports grew by over 30 per cent and continued to grow at 15 per cent per annum until 1975. Simultaneously, Romania joined the IMF and the World Bank and became an active borrower. It appears, therefore, that Romania engaged in a second bout of import-led growth and was encouraged to do this by the emergence of new borrowing opportunities, a hypothesis that is strengthened by Montias's econometric studies which indicate that Romania's decisions to import machinery and equipment from the West depend critically on its short-term capacity to acquire hard currency (Montias, 1980).

The problems of repayments were further complicated by the energy crisis and are discussed in the next chapter.

Notes

1. This is a slightly idiosyncratic version of the product-cycle model. See also Hanson (1981a) and Johnson (1975).
2. The text of Sik's broadcasts is published in Sik (1971).
3. This is probably an underestimate. For further details of calculations see Wiles and Smith (1978).

11 THE ENERGY CRISIS AND EAST-WEST TRADE

The Effects of the Energy Crisis on Eastern and Western Economic Systems

The increase in world oil prices at the end of 1973 had both macroeconomic and microeconomic effects on the import-led growth strategy.

1. *Macroeconomic effects* — the deterioration in the terms of trade of net energy importers in both Eastern and Western Europe resulted in balance of payments pressures, a decline in the resources available to the domestic economy, and led to a deceleration in the rate of growth of living standards and to inflationary pressures. Differences in the adjustment process of the market economies and the centrally-planned economies affected East-West trade directly.
2. *Microeconomic effects* — the increased price of energy and energy-intensive products required a reappraisal of investment programmes and changes in the structure of production and exports but this was further complicated by the introduction of the CMEA sliding world average price system which created a substantial divergence between intra-CMEA prices and world market prices.

Unfortunately for the East European countries that were net energy importers (Hungary, Bulgaria, Czechoslovakia and the GDR), the full impact of the increase in world oil prices occurred just at the time when they should have been bringing their import and export patterns on to a stable long-term trend, but the macro- and microeconomic problems combined to aggravate the problem of external balance.

In any economic system the restoration of internal and external equilibrium, following a deterioration in the terms of trade leading to reduced domestic resource availability, requires either an increase in the growth rate of domestic output (through increased efficiency and productivity) and/or a reduction in the growth of domestic demand.

The (theoretical) transmission mechanism for this process in a market economy is that increased import costs are fed through directly to manufacturers who attempt either to pass these costs on to consumers through price rises and/or to substitute other sources of supply. In a gold-backed system the process would be accompanied by an outflow of gold and an automatic reduction in the domestic money supply which would cause domestic demand to be reduced to the available level of domestic resources at constant prices. Under the conditions prevailing in 1973 with floating (but central bank controlled) exchange rates, the external adjustment process in the face of reduced resource availability required either an acceptance of inflation as a means of restoring domestic equilibrium and/or government measures to reduce the level of domestic demand, including a squeeze on the money supply. In the market economies combinations of the two policies have been used with differing degrees of success, but as inflationary pressures stimulated defensive reactions by those faced with reduced real incomes, and as the monetarist ethos started to prevail, greater emphasis was placed on deflation as a method of restoring equilibrium. A major drawback to deflation is that damping-down on aggregate demand stimulates multiplier effects on industries not affected by the initial price increases, culminating in recession whose effect is most severe in the investment and producer goods industries.

In the centrally-planned economies the *Preisausgleich* prevents external cost pressures from feeding directly into domestic costs, while centralised control over prices and wages prevents them from increasing automatically. There is, however, reduced availability of resources to the economy and either domestic demand must be cut back, or domestic supply increased in some fashion. If consumption is to be cut, equilibrium can only be restored by the processes analysed in Chapter 6. The failure of domestic prices to reflect changed world market conditions provides little incentive for enterprises to cut back on imported inputs, while the cult of gross output provides an incentive to maintain high levels of output which is transmitted throughout the production hierarchy, from the ministry down to the enterprise.

Central authorities also have good reason to avoid cuts in money wage rates and price increases, and have also exhibited a reluctance to cut back on investment projects which they consider will, in the longer run, contribute to a faster output growth which will enable demand to be satisfied. Consequently, the East European economies contained built-in pressure to maintain a high level of aggregate demand in the face of the energy crisis, which could not be sustained from domestic sources.

A short-term complementarity arose between the market economies of Western Europe and the planned economies of Eastern Europe. The governments of the market economies attempted to deflate demand while the capacity of the new wealth holders in the Middle East to absorb western goods was insufficient to maintain output at full employment levels. The oil exporting countries placed their surplus funds with western banks but western manufacturers were unwilling to take up loans to invest in new plant at a time of stagnating demand.

Western manufacturers, particularly in the investment goods industry, sought new markets for their products, while the East European economies attempted to maintain a higher level of aggregate demand than could be sustained from domestic sources. Western banks sought new outlets for their funds and were frequently encouraged by West European governments to lend them to East European countries to facilitate West-East trade.

This systemic complementarity of interests would inevitably be replaced by a competitive situation when the East European economies attempted to sell the outputs of their newly constructed plants in the West, and simultaneously competed for imported raw materials and energy sources.

The *Preisausgleich* and the separation of domestic producers from foreign suppliers and markets created difficulties in identifying which industries to expand and which to contract, resulting in substantial delays in response to changed world market conditions.

The East European countries continued a rapid expansion of their domestic investment programmes in the mid-1970s, largely on the basis of imported machinery and equipment, simultaneously increasing their requirements for imported components and raw materials and incurring greater volumes of indebtedness. As long as western financial confidence was maintained and capital inflows preserved, the policy of continued investment, including investments which themselves required further hard currency imports, could continue. Once confidence started to decline however and problems of repayments arose, the need to cut back on imports to restore balance of payments equilibrium created such effects as domestic bottlenecks and shortages of key materials, which in turn led to falls in output. The later the financial crisis was delayed and the greater the volume of 'uneconomic' investment undertaken, the greater would be the scale of any subsequent adjustment process.

The East European economies were not affected equally by the oil crisis and western bankers who had become more suspicious of the ability of the East European economies to repay credits, and of the willingness of the USSR to assist East European economies that were in difficulties, started to appraise the prospects of the East European countries separately.

Indebtedness, Centralisation and Oil Imports

The use of administrative measures to bring about external equilibrium may be expected to be successful in the short run at the expense of the long run, while decentralising reforms may aggravate balance of payments pressures in the short term.

In the short term, centralisation of foreign trade decisions could help to restore external balance but could be accompanied by domestic bottlenecks which would lead to the rundown of stocks, failure to meet output targets, and so on, which would subsequently result in a rapid increase in imports and a failure to meet export targets.

The use of price criteria alone is unlikely to bring about such a rapid improvement in the external balance, but in the longer run the introduction of prices that reflect the costs of imported components and raw materials and domestic resources in the process of production and reflect export earnings should enable planners to undertake investments that will provide better rates of return to hard currency resources, and will cut the demand for imported components and materials. If enterprise indicators are linked to these prices, enterprises will in the long run be encouraged to economise on high-cost, low-return imports. In the short run, the use of monetary incentives requires an adequate flow of consumer goods to be effective, which may be facilitated by an open policy towards imports, while the experience of Poland in 1972–5, Romania in 1978, and Hungary after 1968 is that decentralising reforms have been accompanied by increases in wage rates.

Finally, a decentralisation of foreign trade decisions, combined with centrally-determined sales or output targets is likely in the short run to lead to an increase in hard currency imports, as enterprises will attempt to import raw materials, components and higher-quality machinery that may not be available from domestic or CMEA sources.

The clearest support for this hypothesis comes from Romania where a stable level of indebtedness was maintained from 1974 to 1976 under the direction of centralised controls, but the continued investment in heavy industry resulted in excess demand for raw materials which saw hard currency debt treble between 1978 and 1980.

Hungary, with an initially high level of indebtedness in the 1970s, following the introduction of the New Economic Mechanism, was forced into some measures of recentralisation by the oil crisis and did not introduce increased energy prices directly into wholesale and retail prices (see Chapter 7). Hungary still remained the most decentralised economy in the bloc (albeit with enterprises with considerable monopoly powers), and imports, exports and indebtedness grew faster than elsewhere in the bloc in the period from 1974 to 1979.[1] Whether decentralisation will bring about a subsequent improvement in the balance of payments remains to be seen.

A final variable, the degree to which the East European economies have been forced on to hard currency markets to buy crude oil, has also been crucial in the build-up of indebtedness. Bulgaria and Czechoslovakia have received a far higher proportion of their total energy imports from the USSR throughout the period (see Table 11.1, p. 225), probably as an indication of approval for their centralised domestic policies and support for Soviet-type integration proposals in the CMEA, and have consequently had to spend less hard currency on crude oil imports.

Table 11.1: Energy Imports and Hard Currency Deficits

	% of energy imports from USSR[a]		Crude oil imports from OPEC 1975–8	
	1970	1978	Tons	Million dollars
Bulgaria	85.5	92.6	3.6	330
Czechoslovakia	84.6	87.0	3.4	300
Hungary	69.6	61.5	4.9	479
Poland	92.1	79.3	13.0	1188
Romania	17.8	8.8	35.3	3218

Note: a. Energy imports measured in standard fuel equivalent.

Sources: Figures calculated from J. P. Stern (1982), pp. 22, 23, 53.

Table 11.2: Trade Balances and Indebtedness in East European Trade With Industrial Market Economies

	CMEA data (million transferable roubles)						Western data (million US dollars)				
	1971		1975		1979		Net indebtedness			Gross debt as % total annual revenue	
	Exports	Imports	Exports	Imports	Exports	Imports	1971	1975	1979	1972	1979
Bulgaria	271	321	324	952	850	879	723	2,257	3,730	198	195
Czechoslovakia	762	893	1,157	1,562	1,746	2,279	160	827	3,070	46	112
Hungary	566	774	855	1,257	2,845	3,597	848	2,195	7,320	140	239
Poland	1,040	992	2,423	4,621	3,514	4,632	764	7,381	20,000	87	333
Romania	644	748	1,255	1,665	2,484	2,637	1,227	2,449	6,700	99	130

Sources: CMEA data from *Statisticheskii Yezhegodnik Stran-Chlenov SEV*, various years. Western data from CIA, National Foreign Assessment Center. Net indebtedness reprinted in Comecon Foreign Trade Data, 1980, pp. 473–4. Gross debt as a percentage of total revenue from Portes (1981) p. 47.

Western data also indicate that the increase in Bulgarian exports to the West in the late 1970s and early 1980s can be attributed mainly to oil and refined oil products. This effectively amounts to a Soviet hard currency subsidy, as the oil is initially imported from the USSR and paid for in transferable roubles at the intra-CMEA price.

Hungary, while having to seek a greater proportion of its energy imports from non-Soviet sources, has been more successful in containing its growth of domestic energy consumption in general, and oil imports in particular. Romania and Poland have both substantially increased crude oil imports, the reasons for which will be examined in the next section.

The Causes of Indebtedness in Poland and Romania

There were a number of broad similarities between the conditions in Poland and Romania. Both countries had ambitious leaders who, with very different personalities, had risen through the domestic Party apparatus, owed little personal allegiance to Moscow and were willing to push through strategies that involved considerable risk. Both leaders were also reputed to be more concerned with the broader political implications of economic strategy than with routine matters of economic administration and calculation, and both tended to centralise key economic decisions.

Crucially, however, while both countries imported western machinery and equipment on a far greater scale in relation to GNP than elsewhere in the bloc, planning in terms of maximising gross output, and in particular planning by material balances with insufficient attention paid to the export value of domestic materials utilised in production, resulted in both countries concentrating their investment in energy-intensive industries, which transformed them from being net energy exporters at the time of the first major world oil price increase to net energy importers by the time of the second increase in 1979. Elsewhere in the bloc the lack of domestic energy may have acted as a constraint to investment on such a scale.

Romania[2]

Part of the original Romanian strategy was that it should provide machinery and equipment (and expertise) to the oil-producing countries and receive crude oil in exchange. Romania expanded its domestic oil refining capacity to 33 million tons per annum (although domestic crude oil production had stabilised at about 14.5 million tons in the mid-1970s), expanded its petrochemical industry and embarked on a somewhat optimistic exploration programme for oil in the Black Sea. It was anticipated that similar deals would enable Romania to obtain raw materials from Third World countries in exchange for machinery and equipment, or enable Romania to earn hard currency from the sale of machinery. The Third World was effectively to take over the role played by the USSR in the CMEA as a supplier of hard goods (energy, raw materials and currency) in exchange for soft goods (machinery and equipment, consumer goods, and so on). The raw materials obtained in this way would be processed on machinery bought from the West, and exported for hard currency sales.

The policy appeared to be broadly successful in the mid-1970s, but many of the economic assumptions on which it was based had been invalidated by the first increase in oil prices, while many of the political assumptions were also questionable. As western manufacturers attempted to pass the increased cost of crude oil on to the consumer, demand for oil products and chemicals fell, margins tightened and the rate of return on investment in these industries declined. The world recession led to a similar decline in demand for such goods as steel and machine tools. The East European producers, who were mainly marginal suppliers to West European markets and did not possess much marketing experience or control over sales outlets, and tended to receive a lower price for sales of seemingly homogeneous chemical and plastic products (OECD, 1980), were likely to be more seriously affected by any downturn in the world market than their competitors.

At a time when western manufacturers were closing existing plant in the steel and petrochemical industries, Romania was expanding its ambitious investment programme in just those industries. In the 1976–80 five-year plan 36.3 per cent of GNP was devoted to investment, half of which was concentrated in industry. Of industrial investment, 85.4 per cent (equivalent to 15.3 per cent of GNP) was concentrated in producers' goods (Group A), and 54.7 per cent of industrial investment (8 per cent of GNP) was concentrated in ferrous and non-ferrous metallurgy, chemicals and machine tools (*Anuarul Statistic*, 1981). Electricity production was planned to grow by 47 per cent to meet industrial needs, but this increase was to be achieved by substantially increasing the proportion of electricity to be generated by coal and low-quality lignite (whose output was planned to double) so that the consumption of oil and natural gas for electricity generation would decline, freeing them for use in the petrochemical industry.

In retrospect, it can be seen that just about every assumption on which this strategy was based proved to be overoptimistic, resulting in substantial hard currency deficits from 1978 onwards. Domestic oil production declined from 14.5 million tons to 11.5 million tons after 1977, instead of rising to a planned 15.5 million tons. Labour unrest in the mining areas prevented the output of hard coal and lignite from rising substantially above the 1975 levels, and despite the diversion of increasing quantities of oil and natural gas to electricity generation, electricity output targets in the five-year plan were underfulfilled by 10–15 per cent. The consequences were that petrochemical output, including plastics and fertilisers, stagnated after 1977, there was a progressive decline in the degree of fulfilment of industrial targets, and agricultural output in the early 1980s remained at or about the 1976 level. Externally, the ousting of the Shah of Iran led to the cancellation of a bilateral deal whereby Romania was to receive 4 million tons of crude oil per annum in exchange for deliveries of machinery and equipment.

The brunt of plan failure was passed on to the foreign trade sector in 1978 and 1979. Industrial investment still expanded by 20 per cent in 1978, the growth being entirely attributable to machinery imports, the deficit on machinery trade going above 1 billion dollars. Declining oil production and expanding demand resulted in Romania becoming a net oil importer (in value terms) in 1977, the collapse of bilateral deals meant that this was paid for in hard currency, and surpluses with Third World countries were turned into deficits from 1977 onwards. By 1979, when world oil prices again doubled, Romania was importing 14 million tons of crude oil at a cost of 2 billion dollars in hard currency. In 1980 the cost of oil imports rose to 3.8 billion dollars, resulting in a deficit on the oil account of 1.6 billion dollars. Simultaneously, Romania imported iron ore from hard currency sources (principally the USA) to maintain steel production. By 1980, despite a cutback in machinery imports, the trade deficit with non-socialist countries was in the region of 1.25 billion dollars, no longer attributable to machinery and equipment imports, but to imports of energy and raw materials to supply the new plants.

The critical feature of Romania's indebtedness has been its isolation from Soviet supplies of crude oil. Romanian hard currency imports of crude oil amounted to 9.6 billion dollars in the 1976–80 plan, largely obtained from the Middle East, to whom Romanian exports amounted to only 2.5 billion dollars over the same period. Simultaneously, Romania's hard currency debt rose from 2.5 billion dollars to 9 billion dollars. In addition to paying for these imports in hard currency Romania did not benefit from the intra-CMEA price subsidy. (The small amounts of Romanian imports from the USSR were paid for in hard currency at world market prices.) Had Romania received all her oil at the intra-CMEA price the subsidy would have amounted to 3.6 billion dollars from 1976 to 1980 (2 billion dollars in 1980 alone). Any other percentage can be prorated accordingly.

Poland[3]

Poland's indebtedness arose from a combination of systemic economic factors, social factors and economic mismanagement, which have caused the dissipation of its natural wealth. Poland is one of the best endowed countries of Eastern Europe with reserves of coal, copper, zinc, silver and sulphur, as a result of which its export prices improved faster than its import prices until 1979.

Poland's import-led growth strategy differed from that embraced elsewhere in Eastern Europe, not only in the degree to which machinery and equipment were imported from the West, but in that from the outset (1972) it extended to imported consumer goods, raw materials, components, and so on. This enabled domestic wage rates and incomes, consumption and investment to grow faster than 10 per cent per annum from 1972 to 1975. The justification for financing current consumption out of borrowing was that this would stimulate increased labour productivity which would in the long run result in increased output in exactly the same way as capital imports, the sale of which would facilitate the repayment of loans (consumption-led growth). In reality, the process only stimulated unsustainable expectations about the future rate of growth of consumption, while planners and Party officials also harboured unrealistic expectations about output growth, as industrial output was boosted in the short-term by taking up capacity that had been underutilised in the Gomulka era (Gomulka, 1982a).

Over one third of GNP was devoted to investment between 1973 and 1976 and much of this investment was concentrated in sectors such as steel, petrochemicals, car production and shipbuilding for which world demand was stagnating, and which required imported raw materials. The pattern of investment displayed many of the classic problems of a centrally-planned economy, but were aggravated by specifically Polish factors.

In addition to the ambitious centrally-determined investment demands, enterprises successfully overindented for capital resources resulting in overfulfilment of capital plans. Excessive capital accumulation resulted in 'investment scatter' — too many projects were started and completion dates fell behind schedule. Resources frozen in incompleted investments amounted to 160 per cent of the planned level of total investment in 1980, while 61 per cent of investments completed in 1980 had exceeded their planned gestation period (Nuti, 1981a, pp. 13–14). This waste of resources was compounded by the fact that delayed completion dates frequently meant that the resulting outputs were obsolete and unsaleable on western markets, while high rates of interest were being paid on the capital borrowed.

The costs of investments and their impact on the balance of payments were inadequately appraised — the increased demands of the construction industry required increased output of the energy-intensive cement industry, in turn worsening the balance of trade in energy and requiring imports of oil from hard currency sources; buses produced under licence from Berliet (France) each contained 6,000 dollars of imported materials and were unsuitable for Polish conditions (Nuti, 1981a).

The problems were complicated by weak and unco-ordinated planning as the centre frequently lost control of the demands of enterprises — bottlenecks in the supply of electricity and other aspects of infrastructure were not foreseen, resulting in power cuts, shortages of inputs, and so on, which were frequently corrected by resorting to imports.

Far greater reliance was placed on the construction of complete installations in greenfield sites than elsewhere in Eastern Europe, with under 25 per cent of industrial investment devoted to modernising existing plant (Nuti, 1981a). The neglect of investment in infrastructure, housing, schooling, hospitals, and so on, made it difficult to attract labour to these sites, while the desire to avoid frictional unemployment made planners reluctant to close down the old unmodernised plant. Consequently, overmanning of old plant coexisted with undermanning of new plant.

Economic maladministration, particularly in the selection of investment, was increased by the promotion of cadres loyal to Gierek to key positions, resulting in incompetence, negligence and corruption (Portes, 1981). Political bias was also shown in the import of agricultural machinery and other inputs which were more suited to the large less-efficient state farms than to the smaller private farms which met much of the domestic demand for food.

The bulk of the impact of mismanagement was borne by the foreign trade sector until 1977, when hard currency constraints necessitated a stabilisation in the level of imports from the West (in constant prices), while attempts were made to boost exports and to seek two to three year credits on imports of semi-manufactures (against a normal six months). The level of investment was cut back and bottlenecks in the economy led to production stoppages. In 1979 national income declined by 2.3 per cent.

Indebtedness to the West grew from 0.8 billion dollars at the end of 1971 to 7.4 billion dollars at the end of 1975 to 20 billion dollars at the end of 1979, when net debt was 16.8 per cent of GNP, three times annual export earnings, and the debt service ratio was 75 per cent according to Polish estimates and 92 per cent according to western estimates. The effect of the two to three year credits raised in 1978 resulted in an impossible level of repayments falling due in 1981 (Portes, 1981, pp. 10–12). Following meetings with representatives of western banks in spring 1980, Polish officials proposed a 27 per cent increase in hard currency earnings for that year, while imports were to decline by 3 per cent (14 per cent in real terms). The planned growth of exports necessitated increased exports of high-quality foodstuffs. The increase in prices of these items, to deter domestic consumption, triggered the strikes, the growth of Solidarity, and the progressive economic decline and virtual collapse of 1981.

Conclusion

Do the Polish and Romanian crises provide any lessons about the operation of centrally-planned economies dependent on foreign trade that can be generalised to the rest of the bloc? Three questions need to be examined:

(1) Why did imports continue to expand until the late 1970s?

(2) Why were those countries unable to generate a sufficient volume

of hard currency exports to pay for their imports?
(3) What was the role of financial factors, and in particular why were continued western loans available?

Imports

The reasons for the expansion of imports may be summarised as follows.

The East European countries embarked on a deliberate attempt to increase domestic economic growth through the acquisition of western technology. Major differences appeared in the speed with, and degree to which, this policy was embraced. Romania effectively initiated the policy in 1967 and after some initial disillusionment resumed the policy in 1972, but Poland adopted the fastest growth from 1972 onwards. Czechoslovakia and Hungary adopted more prudent importing policies, while Hungary placed greater emphasis on imports to raise agricultural productivity and to bring domestic producers into contact with foreign consumers and prices. In most countries, however, the secondary effects of industrial growth in the demand for energy and raw materials were underestimated, as were the multiplier effects of the growth of wages on the demand for foodstuffs and consumer goods. This was most severe in Poland where from 1976 to 1980 73 per cent of credits were used to finance imports of raw materials, energy and food. Disequilibria resulting from overambitious planning were manifested in the foreign trade sector from 1974 to 1977. When a financial crisis arose in 1980, equally ambitious attempts to correct external disequilibrium resulted in internal unrest and virtual economic collapse.

In Romania, disequilibria from overambitious planning were initially borne in the domestic consumer sector. Domestic unrest resulted in some of the impact being transferred to the foreign trade sector from the middle of 1977 onwards, culminating in a financial crisis in 1981. Both countries overstimulated their demand for imported crude oil in the period from 1974 to 1979 and were severely affected by the 1979 increase in world oil prices. Poland and Romania are now both heavily dependent on imported raw materials to keep existing plant (much of which may have to be written off) in operation. This problem is less severe for those countries which pursued a slower rate of expansion,

Romania has recently announced its intention to establish a direct link between international and domestic prices in 1980, by establishing a single rate of exchange between the domestic leu and the transferable rouble and the dollar, which would be used to feed the costs of imported raw materials directly into enterprise accounts. This amounts to a virtual abolition of the *Preisausgleich* at enterprise level, and would enable central authorities to identify loss-making plants. There remains some scepticism in the West as to how far Romanian authorities will be willing to implement this reform and in mid-1980 it was apparent that many enterprises were still receiving budget subsidies to cover losses (Smith, 1981a).

Exports

Holzman (1979) has provided sophisticated analyses to indicate that the East European economies are systemically biased against producing manufactured goods with the quality and performance specifications necessary for volume sales on world markets.

The traditional emphasis on gross output and the disregard of the needs of the consumer have prevented the development of an ethos in which manufacturers respond quickly to changes in consumer demand, which in turn prevents them from competing successfully with western manufacturers in western markets on such factors as quality of output, timing of deliveries, packaging, and consumer fads.

Holzman also argues that attempts to close technology gaps by importing western machinery and equipment have been largely unsuccessful as the technology transferred has frequently represented a 'tried and trusted' but outdated product or process, while the original manufacturer is researching and developing new techniques. Although individual East European countries may excel in certain areas of production, delays in identifying areas for development and in constructing new plant and bringing it into operation increase the time lags before the product is brought into large-scale series production.

By that time prices on world markets may have dropped substantially (if the product is still saleable), multinationals may already be producing competing lines in Third World countries with cheaper labour sources, and attempts to compete on grounds of price are frequently met by western allegations of dumping. All of these factors accompanied Romanian attempts to move into the world textile industries, and are likely to be more acute with belated attempts to move into microelectronics.

Complaints of this kind were made in Romanian trade and

economic journals in the early 1970s.[4] Romanian self-criticism has included complaints that sales personnel do not travel enough to make satisfactory contacts, to develop sales campaigns and display techniques. On the domestic production side criticism is of the 'systemic' kind, citing the remoteness of producers from consumers, poor-quality products, lack of incentives for manufacturers to meet export orders, low-quality intermediate goods and 'storming' preventing exporters from receiving inputs of the necessary quality at the necessary time.

It is difficult to test the hypothesis that the East European countries are systemically incapable of selling basic manufactured commodities to the industrial West due to the large number of exogenous factors which occurred at the time that exports were intended to grow, including absolute and relative price changes and their impact on western imports, western recession, quantitative restrictions, and so on. It is also difficult to propose what should constitute a satisfactory growth and structure of exports for the purposes of comparison. Some of the growth rates of East European exports appear to be impressive. Total East European exports to the West grew 3.6 times from 1970 to 1979 in real terms, exports of manufactures grew four-fold and machinery five-fold, but all from low initial starting points and their share of total western imports declined from 1.5 per cent to 1.4 per cent. Soviet exports (excluding gold) increased their share of western imports from 1.0 per cent to 1.4 per cent[5] and Bulgaria's spectacular growth of exports at the end of the decade was effectively achieved by recycling Soviet oil.

Finance

Finally it must be asked why western loans were continued, after doubt had been cast on the ability of certain East European countries to repay existing loans. One explanation may be that individual participants on both sides continued with a process they knew to be unsustainable in the long run in the hope of minimising the losses to be incurred, or of extricating themselves before the inevitable breakdown occurred.

In retrospect, it can be seen that financial errors and misunderstandings of the partners' modus operandi were made by East and West alike. Romania and Poland paid insufficient attention to the question of the timing of capital repayments and interest payments, leading to unnecessarily difficult cash problems in the early 1980s.

Romania, for example, was scheduled to repay over 4 billion dollars in 1982. Hopes that western inflation would substantially reduce the real value of repayments were severely affected by monetarist policies in the West, which pushed up international interest rates when new loans and refinancing were required. East European analyses also showed a naive assessment of the role of interest groups in western society, and argued that 'monopoly capitalists' who had helped them raise loans in the West would also negotiate the removal of tariffs, quotas, and so on, in order that exports could be made to facilitate loan repayment.

On the western side the fact that loans were (CMEA) government guaranteed may have made them appear less risky than loans to a western corporation that could go bankrupt. Portes (1981, p. 9) provides evidence of western banks' faith in the ability of the Polish government to sell minerals to repay loans as late as 1979. It seems, therefore, that western banks placed less emphasis on the viability of individual projects when assessing loans to Eastern Europe than they would when making loans to individual western companies.

Excessive reliance may have been placed on Soviet ability and willingness to support any bad loans. CIA estimates of Soviet gold reserves of 1,500 tons would, at the high price of 500 dollars an ounce, yield only 25 billion dollars. Similarly, annual production of 300 tons a year would yield 5 billion dollars and my own estimates of Soviet sales of diamonds and precious metals other than gold indicate that these are currently running at about 1.5 billion dollars a year. Other major Soviet hard currency earners are oil and oil products (10–12 billion dollars), natural gas and arms (possibly 5 billion dollars). Soviet import requirements, particularly for grain, must be met from this amount.

It appears unlikely that the USSR has the capability to repay East European indebtedness of over 60 billion dollars, as well as providing Eastern Europe with energy imports. Although western estimates differ substantially in their degree of pessimism concerning Soviet potential for continued oil exports to the West, there is considerable agreement that by the end of the decade natural gas will take over from oil as the principal hard currency earner.

The prospect both for repayments of outstanding debts and future loans to Eastern Europe may depend critically in the long run on the potential of Soviet exports of natural gas. In the shorter run the willingness of the USSR to help East European countries make immediate repayments will depend on their long-term assessment of the prospects of eliminating the problem.

Notes

1. For a discussion of the impact of the energy crisis on the Hungarian reforms see Csikos-Nagy (1978) and Radice (1979).
2. Based on Smith (1979a, 1981b, 1982).
3. Largely based on Portes (1981), Nuti (1981a, 1981b), Gomulka (1978b, 1982a).
4. See Smith (1979a), Burtica (1970, 1971, 1972).
5. Derived from western data in Vienna Institute (1980).

REFERENCES

Abonyi, A (1981) 'Imported Technology, Hungarian Industrial Development and Factors Impeding The Emergence of Innovative Capacity', in Hare *et al.* (eds).

Academia (1958) *Dezvoltarea Economiei RPR pe Drumul Socialismului 1948–57.* Academia Republicii Populare Romine, Bucharest.

Academia (1964) *Dezvoltarea Economica A Rominiei 1944–64.* Academia RPR.

Alampiev, P; Bogomolov, O; Shiryaev, Yu (1973) *A New Approach to Economic Integration.* Progress Publishers, Moscow.

Altmann, HL (1980) 'Czechoslovakia; Prospects For the 1980's', in NATO, 1980.

Amann, R; Cooper, J: Davies, RW (eds) (1977) *The Technological Level of Soviet Industry.* Yale University Press, New Haven.

Andrle, V (1976) *Managerial Power in the Soviet Union.* Saxon House, London.

Anuarul Statistic al Republicii Socialiste Romania. Bucharest (published annually).

Ascherson, N (1981) *The Polish August.* Penguin, Harmondsworth.

Ausch, S (1972) *Theory and Practice of CMEA Co-operation.* Akademiai Kiado, Budapest.

Ausch, S; Bartha, F (1968) 'Theoretical Problems Relating to Prices in Trade between Comecon Countries'. *Soviet and East European Foreign Trade, 4.* Original in *Kozgazdasagi Szemle*, 1967, no. 3.

Balassa, BA (1959) *The Hungarian Experience in Economic Planning.* Yale University Press, New Haven.

Balawyder, A (1980) *Co-operative Movements in Eastern Europe.* Macmillan, London.

Bartha, F (1976) 'Some Ideas on the Creation of a Multilateral Clearing System among the Comecon Countries'. *Soviet and East European Foreign Trade, 12.* Original in *Kulgazdasag* 1975, no. 1.

Baykov, A (1946) *Soviet Foreign Trade.* Princeton University Press, Princeton.

Berend, I (1971) 'The Problem of East European Integration in Historical Perspective' in Vajda and Simai (eds).

Berliner, J (1957) *Factory and Manager in the USSR.* Harvard University Press, Cambridge, Mass.

Bethkenhagen, J (1977) 'Joint Energy Projects and their Influence on Future Comecon Energy Autarchy Ambitions' in NATO, 1977.

Birman, I (1978) 'From the Achieved Level'. *Soviet Studies, 30.*

Birman, I (1980a) 'The Financial Crisis in the USSR'. *Soviet Studies, 32.*

Birman, I (1980b) 'A Reply to Professor Pickersgill'. *Soviet Studies, 32.*

Bogomolov, O (1967) *Mirovaia Ekonomika i Mezhdunarodnye Otnoshenia*, no. 5, Moscow.

Bogomolov, O (1973) 'The International Market of the CMEA Countries' in Kiss (ed.).

Boltho, A (1971) *Foreign Trade Criteria in Socialist Economies.* Cambridge University Press, London.

Bornstein, M (1972) 'Soviet Price Statistics' in Treml and Hardt (eds).

Brabant, J M van (1973) *Bilateralism and Structural Bilateralism in Intra CMEA Trade.* Rotterdam University Press, Rotterdam.

Brabant, J M van (1980) *Socialist Economic Integration.* Cambridge University Press, London.

Brus, W (1972) *The Market in a Socialist Economy* (English translation of work first published in Poland in 1961). Routledge and Kegan Paul, London.

Brus, W (1975) *Socialist Ownership and Political Systems.* Routledge and Kegan Paul, London.

Brus, W (1979) 'The Eastern European Reforms: What Happened to Them?' *Soviet Studies, 31.*

Brus, W (1980) Review of J Kosta 'Abriss der Sozialekonomischen Entwicklung der Tschechoslowakei 1945–1977'. *Soviet Studies, 32.*

Brus, W (1981) 'Economic Reforms as an Issue in Soviet-East European Relations' in Dawisha and Hanson (eds).

Bukharin, N; Preobrazhensky, E (1970) *The ABC of Communism*. Penguin, Harmondsworth.

Burtica, C (1970) *Probleme Economice No. 2*. Bucharest.

Burtica, C (1971) *Probleme Economice No. 7*. Bucharest.

Burtica, C (1972) *Probleme Economice No. 4*. Bucharest.

Bush, K (1973) 'Soviet Inflation' in NATO (1973).

Chvojka, P (1977) 'International Monetary Relations in the Development of Comecon'. *Soviet and East European Foreign Trade, 13* (Original in *Politicka Ekonomie*, 1976 no. 11).

CIA (1979) National Foreign Assessment Center. *Handbook of Economic Statistics.*

CMEA (1977) *The Multilateral Economic Co-operation of Socialist States: A Collection of Documents*. Progress Publishers, Moscow.

CMEA Statistics. *Statisticheskii Yezhegodnik Stran-Chlenov SEV*. Moscow. Annual publication.

Cohen, S (1980) *Bukharin and the Bolshevik Revolution*. Oxford University Press (paperback edition), Oxford.

Csikos-Nagy, B (1978) 'The Hungarian Reform after Ten Years', *Soviet Studies, 30.*

Davies, RW (1965) 'Planning for Rapid Growth in the USSR'. *Economics of Planning, 5.*

Davies, RW (1969) 'A Note on Soviet Grain Statistics'. *Soviet Studies, 21.*

Davies, RW (1980) *The Socialist Offensive. The Collectivisation of Agriculture 1929–30.* Macmillan, London.

Dawisha, K; Hanson, P (eds) (1981) *Soviet-East European Dilemmas*. Heinemann, London.

Day, RB (1975) 'Preobrazhensky and the Theory of the Transition Period'. *Soviet Studies, 27.*

Denton, GR (ed.) (1971) *Economic Integration in Europe*. Weidenfeld and Nicholson, London.

Dobb, M (1966) *Soviet Economic Development since 1917*. 6th edition. Routledge and Kegan Paul, London.

Dorner, P (1972) *Land Reform and Economic Development*. Penguin, Harmondsworth.

Drewnowski, J (1961) 'The Economic Theory of Socialism'. *Journal of Political Economy, 69.*

Drewnowski, J (1979) 'The Central Planning Office on Trial: An Account of the Beginnings of Stalinism in Poland'. *Soviet Studies, 31.*

Drewnowski, J (ed.) (1982) *Crisis in the East European Economy*. Croom Helm, London.

Dunning (1969) *Foreign Capital and Economic Growth in Europe* in Denton (ed.).

Elkan, W (1973) *An Introduction to Development Economics*. Penguin, Harmondsworth.

Ellman, MJ (1968) 'Optimal Planning'. *Soviet Studies, 20.*

Ellman, MJ (1971) *Soviet Planning Today*. Cambridge University Press, Cambridge.

Ellman, MJ (1973) *Planning Problems in the USSR*. Cambridge University Press, Cambridge.

Ellman, MJ (1975) 'Did the Agricultural Surplus Provide the Resources for the Increase in Investment in the USSR in the First Five Year Plan?' *Economic Journal, 85.*

Fallenbuchl, ZM (1970) 'The Communist Pattern of Industrialisation'. *Soviet Studies, 21.*

Fallenbuchl, ZM (ed.) (1976) 'Economic Development in the Soviet Union and Eastern Europe'. Vol. 2. Praeger, New York.

Farrell, JP (1975) 'Bank Control on the Wage Fund in Poland: 1950–1970'. *Soviet Studies, 27.*

Feinstein, CH (ed.) (1967) *Socialism, Capitalism and Economic Growth*. Cambridge University Press, Cambridge.

Fejto, F (1974) *A History of the People's Democracies; Eastern Europe since Stalin*. Penguin, Harmondsworth.

Fischer-Galati, S (1967) *The New Romania*. Cambridge, Mass.

Friss, I (ed.) (1978) *Essays on Economic Policy and Planning in Hungary*. Corvina Kiado, Gyoma.

Gabor, I; Galasi, P (1981) 'The Labour Market in Hungary since 1968' in Hare *et al.* (eds).

Garvy, G (1977) *Money, Financial Flows and Credit in the Soviet Union*. Ballinger, Cambridge, Mass.

Gomulka, S (1971) *Inventive Activity, Diffusion and the Stages of Economic Growth*. Institute of Economics, Aarhus.

Gomulka, S (1976) 'Soviet Post-War Industrial Growth, Capital-Labour Substitution and Technical Change' in Fallenbuchl (ed.).

Gomulka, S (1977) 'Economic Factors in the Democratization of Socialism and the Socialization of Capitalism'. *Journal of Comparative Economics, 1.*

Gomulka, S (1978a) 'Slowdown in Soviet Industrial Growth 1947–75 Reconsidered'. *European Economic Review, 10.*

Gomulka, S (1978b) 'Growth and the Import of Technology: Poland 1971–80'. *Cambridge Journal of Economics, 2.*

Gomulka, S (1982a) 'Macroeconomic Reserves, Constraints and Systemic Factors in the Dynamics of the Polish Crisis 1980–82'. Forthcoming in *Jahrbuch der Wirtschaft Osteuropas*. Munich.

Gomulka, S (1982b) 'Kaldors Stylized Facts, Dynamic Economies of Scale and Diffusional Effect in Productivity Growth'. Mimeo, (Unpublished).

Granick, D (1975) *Enterprise Guidance in Eastern Europe*. Princeton University Press, Princeton, NJ.

Gregory, PR; Stuart, RC (1974) *Soviet Economic Structure and Performance*. Harper and Row, New York.

Hanson, P (1981a) *Trade and Technology in Soviet-Western Relations*. Macmillan, London.

Hanson, P (1981b) 'Soviet Trade with Eastern Europe' in Dawisha and Hanson (eds).

Hare, PG; Radice, HK; Swain, N (eds) (1981) *Hungary: A Decade of Economic Reform*. George Allen and Unwin, London.

Hewett, EA (1974) *Foreign Trade Prices in the CMEA*. Cambridge University Press, London.

Hirschman, AO (1975) *The Strategy of Economic Development*. 17th edition. Yale University Press, New Haven.

Hohmann, HH; Kaser M; Thalheim KC (eds) (1975) *The New Economic Systems of Eastern Europe*. Hurst, London.

Holesovsky, V (1977) 'Czechoslovak Economies in the Seventies' in 'United States 1977'.

Holesovsky, V (1980) 'Czechoslovakia: Economic Reforms' in NATO, 1980.

Holzman, FD (1962a) 'Soviet Foreign Trade Pricing and the Question of Discrimination'. *Review of Economics and Statistics, 44.*

Holzman, FD (1962b) 'Soviet Bloc Mutual Discrimination. A Comment'. *Review of Economics and Statistics, 44.*

Holzman, FD (1965) 'More on Soviet Bloc Trade Discrimination'. *Soviet Studies, 17.*
Holzman, FD (1979) 'Some Systemic Factors Contributing to the Convertible Currency Shortages of CPE's'. *American Economic Review, 69.*
IMF 'International Financial Statistics' (monthly).
Ionescu, I (1974) *Revue de l'Est, 5,* Paris.
Jackson, D; Turner, HA: Wilkinson F (1972) *Do Trade Unions Cause Inflation?* Cambridge University Press, Cambridge.
Jahrbuch der Wirtschaft Osteuropas. Osteuropa-Instituts, Munich. Various issues (not strictly published annually).
Jeffries, I (ed.) (1981) *The Industrial Enterprise in Eastern Europe.* Praeger, New York.
Johnson, HG (1975) *Technology and Economic Interdependence.* Macmillan, London.
Jorgensen, D (1971) *Testing Alternative Theories of a Dual Economy* in Livingstone (ed.).
Kaplan, NM: Moorsteen, R (1960) 'An Index of Soviet Industrial Output'. *American Economic Review, 50.*
Kaser, M (1967) *Comecon.* 2nd edition, Oxford University Press, London.
Kaser, M (1981) 'The Industrial Enterprise in Bulgaria' in Jeffries (ed.).
Katsenelinboigen, AI (1975) 'Disguised Inflation in the Soviet Union' in NATO 1975.
Kemeny, G (1952) *Economic Planning in Hungary.* Royal Institute of International Affairs, London.
Kenedi, J (1981) *Do It Yourself: Hungary's Hidden Economy.* Pluto Press, London.
Keren, M (1973) 'The New Economic System in the GDR: An Obituary'. *Soviet Studies, 24.*
Keynes, JM (1940) *How to Pay for the War.* Macmillan and Harcourt Brace, London.
Khrushchev, NS (1962a) 'Vital Questions of the Development of the Socialist World System'. *World Marxist Review, 5.*
Khrushchev, NS (1962b) *Razvitie Ekonomikii SSSR i Partiinoe Rukovodstvo Narodnim Khozyaistvom.* Gospolitizdat, Moscow.
Kirschbaum, JM (1980) 'The Co-operative Movement in Slovakia 1945–1948' in Balawyder (ed.).
Kiss, T (1976) 'International Co-operation in Planning within Comecon'. *Eastern European Economics.* Original in *Kozgazdasagi Szemle*, 1975, no. 6.
Kiss, T (ed.) (1973) *The Market of Socialist Integration.* Akademiai Kiado, Budapest.
Kohler, H (1965) *Economic Integration in the Soviet Bloc.* Praeger, New York.
Kohoutek, M (1968) 'On Problems of the Plan and Market'. *Czechoslovak Economic Papers, 10.*
Kornai, J (1959) *Overcentralization in Economic Administration.* Oxford University Press, London.
Kosygin, AN (1974) *Izbrannye Stati i Rechi.* Politizdat, Moscow.
Kozma, G (1981) 'The Role of the Exchange Rate in Hungary's Adjustment to External Economic Circumstances' in Hare *et al.* (eds).
Kyn, O (1975) 'Czechoslovakia' in Hohmann *et al.* (eds).
Kyn, O; Sekerka, B; Hejl, L (1967) 'A Model for the Planning of Prices' in Feinstein (ed.).
Ladygin, L; Shiryaev, Yu (1966) *Voprosy Ekonomiki,* no. 5. Moscow.
Lazaroiu, D (1967) *Probleme Economice*, no. 9. Bucharest.
League of Nations (1933) *Review of World Trade.* Geneva.
Lebedinskas, AA (1974) *Dengi i Kredit*, no. 12. Moscow.

Lewin, M (1975) *Political Undercurrents in Soviet Economic Debates.* Pluto Press, London.
Livingstone, I (ed.) *Economic Policy for Development.* Penguin, Harmondsworth.
Lufsandorj, P (1978) *Planovoe Khozyaistvo,* no. 9. Moscow.
Lupu, M (1969) *Probleme Economice, 22.* Bucharest.
McMillan, CH (1973) 'Factor Proportions and the Structure of Soviet Foreign Trade'. *ACES Bulletin.*
Madgearu, V (1930) *Rumania's New Economic Policy.* Orchard House, London.
Marczewski, J (1974) *Crisis in Socialist Planning.* Praeger, New York.
Marer, P (1981) 'The Mechanism and Performance of Hungary's Foreign Trade 1968–79' in Hare *et al.* (eds).
Marer, P and Montias, JM (eds) (1980) *East European Integration and East-West Trade.* Indiana University Press, Bloomington.
Marrese M (1981) 'The Evolution of Wage Regulation in Hungary' in Hare *et al.* (eds).
Mendershausen, H (1959) 'The terms of trade between the Soviet Union and smaller countries'. *Review of Economics and Statistics, 41,* 106–18.
Millar, JR (1970) 'Soviet Rapid Development and the Agricultural Surplus Hypothesis'. *Soviet Studies, 22.*
Millar, JR (1974) 'Mass Collectivisation and the Contribution of Agriculture to the First Five Year Plan: A Review Article'. *Slavic Review, 33.*
Montias, JM (1967) *Economic Development in Communist Romania.* MIT Press, Cambridge, Mass.
Montias, JM (1980) 'Romania's Foreign Trade Between East and West' in Marer and Montias (eds).
Montias, JM (1981) 'Observations on Strikes, Riots and Other Disturbances' in Triska and Gati (eds).
Murgescu, C (1967) *'Eficienta Economica a Comertului'.* Academia RPR, Bucharest.
NATO Economics Directorate. Proceedings of Annual Colloquium, Brussels.
1973 — 'Money, Banking and Credit in Eastern Europe'.
1975 — 'Economic Aspects of Life in the USSR'.
1977 — 'Comecon: Progress and Prospects'.
1980 — 'Economic Reforms in Eastern Europe and Prospects for the 1980's'.
Neuberger, E (1966) 'Libermanism, Computopia and the Visible Hand: The Question of Informational Efficiency'. *The American Economic Review, 56.*
Nove, A (1964) *Was Stalin Really Necessary?* George Allen and Unwin, London.
Nove, A (1969) *An Economic History of the USSR.* Penguin, Harmondsworth.
Nove, A (1980) *The Soviet Economic System.* 2nd edition. George Allen and Unwin, London.
Nove, A; Nuti, DM (eds) (1972) *Socialist Economics.* Penguin, Harmondsworth.
Nuti, DM (1981a) 'The Polish Crisis: Economic Factors and Constraints'. *Socialist Register.* Merlin Press, London.
Nuti, DM (1981b) 'Industrial Enterprises in Poland, 1973–80' in Jeffries (ed.).
Nuti, DM (1982) 'The Polish Crisis; Economic Factors and Constraints' in Drewnowski (ed.).
OECD (1980) *East-West Trade in Chemicals.* OECD, Paris.
Ofer, G; Pickersgill, J (1980) 'Soviet Household Saving: A Cross Section Study of Soviet Emigrant Families'. *Quarterly Journal of Economics, 94.*
Pecsi, K (1981) *The Future of Socialist Economic Integration.* M E Sharpe Inc, New York.
Podolski, TM (1973) *Socialist Banking and Monetary Control.* Cambridge University Press, London.
Portes, R (1977) 'The Control of Inflation: Lessons from East European Experience'. *Economica, 44.*

Portes, R (1980) 'Effects of the World Economic Crisis on the East European Economies'. *The World Economy 1.*

Portes, R (1981) The Polish Crisis: Western Economic Policy Options. Royal Institute of International Affairs, London.

Portes, R; Winter, D (1977) 'The Supply of Consumption Goods in Centrally Planned Economies'. *Journal of Comparative Economics, 1.*

Portes, R; Winter, D (1978) 'The Demand for Money and for Consumption Goods in Centrally Planned Economies'. *Review of Economics and Statistics, 60.*

Portes, R; Winter D (1980) 'Disequilibrium Estimates for Consumption Goods Markets in Centrally Planned Economies'. *Review of Economic Studies, 47.*

Preobrazhensky, E (1965) *The New Economics.* Oxford University Press, London.

Promsky, NI (1977) *Izvestiya Nauk; Seriya Ekonomicheskaya No. 4.* Moscow.

Pryor, FL (1963) *The Communist Foreign Trade System.* George Allen and Unwin, London.

Puiu, A (1969) *Probleme Economice No. 12.* Bucharest.

Puiu, A; Ciulea (1967) *Lupta de Clasa No. 8.* Bucharest.

Quigley, J (1974) *The Soviet Foreign Trade Monopoly.* Ohio State University Press.

Radice, HK (1979) 'Csikos-Nagy on the Hungarian Economic Reforms: A Comment'. *Soviet Studies, 31.*

Radice, HK (1981a) 'Industrial Co-operation between Hungary and the West' in Hare *et al.* (eds).

Radice, HK (1981b) 'The State Enterprise in Hungary' in Jeffries (ed.).

Rakowska-Harmstone, T (1976) 'Socialist Internationalism — A New Stage'. *Survey, 22.*

Ribalkin, VE (1978) *Mezhdunarodnii Rinok SEV.* Moscow.

Rosenstein-Rodan, P (1943) 'Problems of Industrialisation of Eastern and South-Eastern Europe'. *Economic Journal, 53.*

Rudcenko, S (1979) 'Household Money Income, Expenditure and Monetary Aspects in Czechoslovakia, GDR, Hungary and Poland 1956–75' in *Jahrbuch der Wirtschaft Osteuropas, 8.* Munich.

Schiavone, G (1981) *The Institutions of Comecon.* Macmillan, London.

Schroeder, G (1975) 'Consumer Goods Availability and Repressed Inflation in the Soviet Union' in NATO, 1975.

Schumpeter, J (1974 edn) *Capitalism, Socialism and Democracy.* George Allen and Unwin, London.

Scinteia, Bucharest. Published daily.

Selucky, R (1972) *Economic Reforms in Eastern Europe.* Praeger, New York.

Shlaim, A; Yannopoulos, GN (eds) (1978) *The EEC and Eastern Europe.* Cambridge University Press, London.

Sik, O (1967) *Plan and Market under Socialism.* Czechoslovak Academy of Sciences, Prague.

Sik, O (1971) *Czechoslovakia: The Bureaucratic Economy.* American edition. IASP, White Plains, New York.

Simes, DK (1975) 'The Soviet Parallel Market', in NATO, 1975.

Smith, AH (1979a) 'Romanian Economic Relations with the EEC'. *Jahrbuch der Wirtschaft Osteuropas, 8*, Munich.

Smith, AH (1979b) 'Plan Co-ordination and Joint Planning in CMEA'. *Journal of Common Market Studies.*

Smith, AH (1980) 'Romanian Economic Reforms' in NATO, 1980.

Smith, AH (1981a) 'The Romanian Industrial Enterprise' in Jeffries (ed.).

Smith, AH (1981b) 'Economic Factors Affecting Soviet-East European Relations in the 1980's' in Dawisha and Hanson (eds).

Smith, AH (1982) 'Is There a Romanian Economic Crisis' in Drewnowski (ed.).

Spulber, N (1957) *The Economics of Communist Eastern Europe.* Wiley, New York.

Statisticka Ročenka C.S.S.R. Prague (published annually).
Stern, JP (1982) *East European Energy and East-West Trade in Energy.* Policy Studies Institute, London.
Sutton, AC (1968, 1971, 1973) *Western Technology and Soviet Economic Development.* Three volumes. Stanford University Press, Stanford.
Tardos, M (1978) 'Relationships Between the International Division of Labour and Hungarian Economic Policy' in Friss (ed.).
Treml, VG: Hardt, JP (eds) (1972) *Soviet Economic Statistics.* Duke University Press, Durham, North Carolina.
Triska, JF; Gati, C (eds) (1981) *Blue-Collar Workers in Eastern Europe.* George Allen and Unwin, London.
Turcan, JR (1977) 'Some Observations on Retail Distribution in Poland'. *Soviet Studies, 29.*
Tyson, Ld'A (1981) 'Aggregate Economic Difficulties and Workers' Welfare' in Triska, JF and Gati, C (eds).
United Nations (1949) *Economic Survey of Europe in 1948.* Economic Commission for Europe, Geneva.
United States (1955) *Trends in Economic Growth.* US Library of Congress.
United States (1977) *East European Economies Post-Helsinki.* US Library of Congress.
Vajda, I, Simai, M (eds) (1971) *Foreign Trade in a Planned Economy.* Cambridge University Press, Cambridge.
Vienna Institute for Comparative Economic Studies (ed.) (1980) *Comecon Data 1979.* Macmillan, London.
Vienna Institute for Comparative Economic Studies (ed.) (1981) *Comecon Foreign Trade Data 1980.* Macmillan, London.
Vincze, I (1978) 'Multilaterality, Transferability and Exchangeability'. Original in *Kozgazdasagai Szemle*, 1978, no. 1.
Vneshnyaya Torgovlya SSSR. Statisticheskii Sbornik. Moscow. Published annually.
Vogel, H (1975) 'Bulgaria' in Hohmann *et al.* (eds).
Volgyes, I (1981) 'Hungary: The Lumpenproletarianization of the Working Class' in Triska and Gati (eds).
Walker, A. (1978) *Marx: His Theory and its Context.* Macmillan, London.
Wheeler, GS (1973) *The Human Face of Socialism.* Lawrence Hill, New York.
Wilczynski, J (1978) *Comparative Monetary Economics.* Macmillan, London.
Wiles, PJD (1962) *The Political Economy of Communism.* Blackwell, Oxford.
Wiles, PJD (1968) *Communist International Economics.* Blackwell, Oxford.
Wiles, PJD (1973) 'On Purely Financial Convertibility' in NATO, 1973.
Wiles, PJD (1974) 'The Control of Inflation in Hungary'. *Economie Appliquée.*
Wiles, PJD (1979) *Economic Institutions Compared.* Blackwell, Oxford.
Wiles, PJD (ed.) (1982) *The New Communist Third World.* Croom Helm, London.
Wiles, PJD; Smith, AH (1978) 'The Convergence of the CMEA on the EEC' in Shlaim and Yannopoulos (eds).
Wiles, PJD; Smith, AH (1982) 'The General View; Especially from Moscow' in Wiles (ed.).
Zauberman, A (1964) *Industrial Progress in Poland, Czechoslovakia and East Germany 1937–1962.* Oxford University Press, London.

APPENDIX

Official Soviet and East European Statistical Publications (published annually)

Bulgaria	Statisticheski Godishnik NRB
Czechoslovakia	Statisticka Ročenka ČSSR
GDR	Statistisches Jahrbuch der DDR
Hungary	Statisztikai Evkonyv
	Statistical Yearbook (in English)
	Statistical Pocketbook of Hungary (in English)
Poland	Rocznik Statystyczny
	Rocznik Statystyczny Handlu Zdgranicnego
	Concise Statistical Yearbook of Poland (in English)
Romania	Anuarul Statistic al RSR
USSR	Narodnoye Khozyaistvo SSSR
	Vneshnyaya Torgovliya SSSR
	SSSR v Tsifrakh
CMEA	Statisticheskii Yezhegodnik stran-chlenov SEV

General (published in Moscow)

Narodnoye Khozyaistvo Sotsialisticheskikh Stran (contains communiques on fulfilment of annual plans)
Mir Sotsialism v Tsifrakh i Faktakh 1980

INDEX

References to East European countries and authors cited in brackets in the text have not been included in the index.